INTERNATIONAL EXPLORATION ECONOMICS, RISK, AND CONTRACT ANALYSIS

INTERNATIONAL EXPLORATION ECONOMICS, RISK, AND CONTRACT ANALYSIS

Daniel Johnston

Disclaimer: The recommendations, advice, descriptions, and the methods in this book are presented solely for educational purposes. The author and publisher assume no liability whatsoever for any loss or damage that results from the use of any of the material in this book. Use of the material in this book is solely at the risk of the user.

Copyright © 2003 by
PennWell Corporation
1421 South Sheridan Road
Tulsa, Oklahoma 74112-6600 USA

800.752.9764
+1.918.831.9421
sales@pennwell.com
www.pennwell-store.com
www.pennwell.com

Marketing Manager: Julie Simmons
National Account Executive: Barbara McGee
Director: Mary McGee
Managing Editor: Marla Patterson
Production/Operations Manager: Traci Huntsman
Managing Editor: Marla M. Patterson
Production Editor: Sue Rhodes Dodd
Book Designer: Beth Caissie
Author Photo: William Gnade

Library of Congress Cataloging-in-Publication Data

Johnston, Daniel.
International exploration economics, risk, and contract analysis/by Daniel Johnston.—1st ed.
p. cm.

ISBN 0-87814-887-6
ISBN13 978-0-87814-887-5

1. Petroleum industry and trade—Finance. 2. Petroleum industry and trade—Management.
3. International business enterprises—Finance. I. Title.
HD9560.5 .J623 2003
622'.338'0681—dc21   2003004332

Printed in the United States of America

4 5 6 7   13 12 11

To Jill

CONTENTS

ACKNOWLEDGMENTS

I would like to express my gratitude to Dr. Ted Coe, Ms. Patti Balentine, and Ms. Dolores Argo at the Institute of Petroleum Accounting. And, in particular I would like to thank Ms. Patti Balentine, my editor at the *Petroleum Accounting and Financial Management Journal (PAFMJ)* with whom I have worked so closely all these years. Thanks for your help Patti.

My PennWell editor, Marla Patterson, is greatly appreciated too. And, my sincere thanks to Sue Rhodes Dodd of Amethyst Enterprises, Tulsa, Oklahoma for her help on this book and so many others over the years.

I would also like to thank my brother David Johnston for his help in organizing and reviewing all this material for me.

INTRODUCTION

For nearly a decade I have written pieces for the *Petroleum Accounting and Financial Management Journal (PAFMJ)* published by the Institute of Petroleum Accounting at the University of North Texas (UNT) in Denton, Texas.

This book is a collection of my articles and columns over the years with additional commentary and material on particular issues.

There is some repetition in the various chapters, which perhaps might be inevitable considering the time frame for these pieces. However, when one considers the areas of repetition, it is understandable—they flow from my nearly constant confrontation with lack of standards in the industry regarding aspects of fiscal system analysis and design. At times it has taken on characteristics of a battle— an uphill battle. However, progress has been made over the past 10 years. The collective understanding in the industry has certainly improved, and individuals are better able to communicate their thoughts and ideas on this subject. Nevertheless, the science of fiscal system analysis and design is far from having standardized terminology. There are still many myths and misconceptions; we have a ways to go.

Chapter 1 (Summer 1994, Vol 13, No. 3)
International Petroleum Fiscal Systems—PSCs

This chapter is a general introduction to the arithmetic and mechanics of the two main types of petroleum fiscal arrangement: royalty/tax systems and production-sharing systems. It also provides basic information about petroleum taxation theory and how it is influenced by the dramatic risk/reward relationships that characterize the petroleum exploration business. Much of this chapter has been heavily updated from its initial 1994 debut. This is partly because of the importance of fundamentals covered in this chapter and also because it was the oldest by far of all my pieces for the *PAFMJ*.

Chapter 2 (Summer 1996, Vol 15, No. 2)
State-of-the-Art in Petroleum Fiscal System Analysis

This chapter is somewhat redundant because of some of the additions and updating in Chapter 1. It is probably significant that such concepts as effective royalty rate (ERR) were first introduced as late as 1996.

Chapter 3 (Summer 1997, Vol 16, No. 2)
Thinking of Going International?—Some Useful Tips

This chapter deals with issues confronting a typical U.S, oilman (or company) considering going overseas for the first time and the differences between domestic vs. international exploration. All too often during the early 1990s, many companies were getting their start in the international sector in Russia or other republics of the Former Soviet Union (FSU)—not a good place to start.

Chapter 4 (Summer 1998, Vol 17, No. 2)
Trends and Issues in Foreign PSCs

This chapter has some of the first examples of "weaknesses" of the "take" statistics. And, it was one of the first introductions of the "entitlement index" concept.

Chapter 5 (Summer 1999, Vol 18, No. 2)
Current Developments in PSCs

International exploration acreage has taken on more of the characteristics of a commodity. The concept of a "balance" between prospectivity and fiscal terms is an old one and predates the petroleum industry. Only the terminology has changed as shown by the summary of Adam Smith's view of rent governed by "fertility" and "situation."

Chapter 6 (Fall/Winter 1999, Vol 18, No. 3)
The International Gas Industry

Stranded gas is a big subject. This chapter deals with some of the reasons why oil is so much more valuable. Gas development options are also discussed.

Chapter 7 (Spring 2000, Vol 19, No. 1)
Key Concerns of Governments and Oil Companies— Alignment of Interests

This chapter focuses on the main concerns of governments designing fiscal terms and awarding acreage: division of profits, division of revenues, savings incentive, maximum efficient production rate, etc. The concept of alignment of interests and how much of that is embodied in the "savings index" is also covered.

Chapter 8 (Summer 2000, Vol 19, No. 2)
Fiscal System Design—The Ideal System

This chapter deals with what in my opinion would be the ideal fiscal system from the perspective of grassroots design. It deals with both allocation strategy as well as fiscal marksmanship.

Chapter 9 (Fall/Winter 2000, Vol 19, No. 3)
Economic Modeling/Auditing—Art and Science, Part I

Every economic model has its weaknesses. Too often management makes investment decisions based on bad numbers. Chapters 9 and 10 were written to address the ever-present "bugs" in cash flow models and how to detect them. The focus of this chapter is to show how to ensure that the model itself is working properly.

Chapter 10 (Spring 2001, Vol 20, No. 1)
Economic Modeling/Auditing— Art and Science, Part II

Chapter 10 examines the assumptions that go into these models and how to review these assumptions quickly to get a feel for whether or not things are realistic and in balance.

Chapter 11 (Summer 2001, Vol 20, No. 2)
Finger on the Pulse—Phuket 2001

This chapter summarizes some key issues that surfaced in my Production Sharing Contracts Roundtable in Phuket, Thailand during the Summer of 2001. In particular, the concept of greed and the important topic of "booking barrels."

Chapter 12 (Fall/Winter 2001, Vol 20, No. 3)
Kashagan and Tengiz

The Kashagan discovery in the Kasakh sector of the North Caspian may hold more recoverable oil reserves than the entire United States. The technical and fiscal/contractual aspects of this discovery is discussed and compared to the nearby giant Tengiz field onshore.

Chapter 13 (Spring 2002, Vol 21, No. 1)
The Bidding Dilemma—a 20-Year Retrospective

For the past two decades, the petroleum exploration industry has suffered from lack of financial success—in other words, "huge losses." One reason is that fiscal terms are too tough. The natural question is: "Why?"

Chapter 14 (Summer 2002, Vol 21, No. 2)
Retrospective, Government Take— Not a Perfect Statistic

The most commonly quoted statistic in the science of petroleum fiscal system analysis is "government take." While it is certainly a useful statistic, it becomes more meaningful when both the strengths and weaknesses are known and understood. This chapter adds more dimension to this subject.

Chapter 15   Additional Commentary on Key Issues

This chapter provides added observations and discussion on a few important topics, such as the value of reserves in the ground, booking barrels, and maximum efficient production rates.

Chapter 16   Example Contracts

A number of example contracts and/or fiscal systems from around the world are summarized in this chapter. The focus is on the commercial terms and a quick summary of government take, effective royalty rate, entitlement, and savings index is provided for each contract.

1

International Petroleum Fiscal Systems—PSCs

The interchangeable use of imprecisely defined terms has given rise to much confusion in this industry. For example, *production-sharing contracts* (PSCs) are frequently called *production-sharing agreements* (PSAs) and lately in some countries *exploration production-sharing agreements* (EPSAs) or *exploration and development production-sharing agreements* (EDPSAs). To refer to the petroleum taxation and contractual arrangements of a country simply as the *fiscal system* is not precisely correct. The practice is common, though, and convenient.

This book adheres to the prevailing terminology that constitutes the language of the industry today. Here, the term *fiscal system* is used somewhat loosely to encompass virtually all taxes, levies, legislative, and contractual aspects of petroleum operations within a sovereign nation/state and its provinces. But a distinction is usually made between elements that constitute the work commitment associated with foreign operations and the attendant royalties, taxes, production sharing, etc.

The host government, which is often represented by either a national oil company or an oil ministry or both, is simply referred to here as the *state* or the *government*. The term *contractor* has specific connotations that are explained later but is used here to mean any company operating in the international arena.

ECONOMIC RENT

The concept of *economic rent* comes from the foundations of economic theory and the produce of the earth, which is derived from labor, machinery, and capital. *Rent theory* deals with how this produce is divided among the holders of the land, the owners of the capital, and the laborers through profit, wages, and rent. A strict distinction can be made between profits and rent, but in the popular language of the industry, this distinction is sometimes missed. But it is an important issue. Excess profits are synonymous with economic rent. That is the way it is defined in this book. However, there are other definitions used in the industry. For example, some economists equate rent with profit.

Economic rent is the difference between the value of production and the cost to extract it. The extraction cost consists of normal exploration, development, and operating costs as well as required rates of return or share of profit for the contractor. Rent deals with the surplus. Governments attempt to capture as much as possible of the economic rent through various means, including taxation, levies, royalties, and bonuses.

The problem for governments in determining how to efficiently capture rent is that nearly 9 out of 10 exploration efforts are not successful. This important element of risk strongly characterizes the upstream end of the oil industry. Developing fiscal terms that are capable of yielding sufficient potential rewards for exploration efforts must account for this risk. It is not an easy matter. Present value theory, expected value (EV) theory, and taxation theory are the foundation stones of fiscal system design and analysis.

The objective of host governments in designing petroleum fiscal systems is to structure an efficient system where exploration and development rights are acquired by those companies who place the highest value on those rights. In an efficient market, competitive bidding can help achieve this objective. But one of the hallmarks of an efficient market is availability of information. Exploration efforts are dominated by numerous unknowns and uncertainty. In the absence of sufficient competition, efficiency must be designed into the fiscal terms.

Governments can seek to capture economic rent at the time of the transfer of rights through signature bonuses or during the production phase of a contract, or concession through royalties, production sharing, or taxes.

Royalties, taxes, and/or production-sharing formulas used for extracting rent are contingent upon production. The contractor and government therefore share in the risk that production may not result from exploration efforts. An important aspect as far as risk is concerned is that oil companies are risk-takers who theoretically diversify their risk. On the other hand, as far as their exposure in the exploration business is concerned, governments are not likely to be diversified. Their risk aversion level is quite different than that of an international oil company. This aspect provides much of the dynamics of international negotiations and fiscal design.

Theoretically, a simple bonus bid with no royalties or taxes would be the ultimate example of a system where the government captured the economic rent at time of transfer. In an efficient market with perfect information and sufficient competition, the bonus would equal the present value of the total expected economic rent. This kind of behavior is seen to some degree in production acquisitions between companies where oil production is purchased and sold. From the government point of view, there is a trade-off between risk aversion (leaning toward bonus bids and royalties to some extent) and risk sharing (through production sharing or profit sharing through taxation schemes).

Figure 1–1 illustrates the basic elements in the allocation of revenues for recovery of costs and the division of profits.

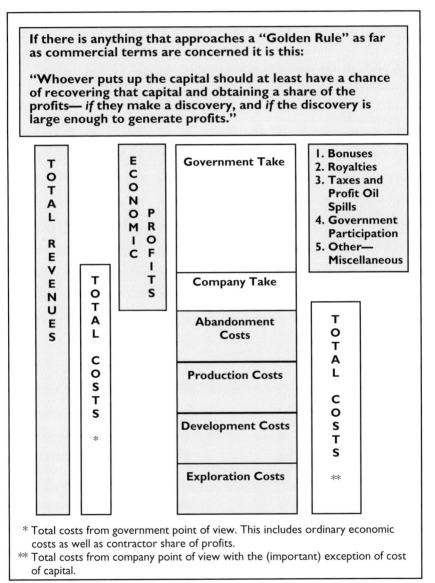

Fig. 1–1 Division of revenues

Governments have devised numerous frameworks for the extraction of economic rents from the petroleum sector. Some are very efficient and some perhaps not. Some are well balanced and cleverly designed and some are quite complex. But the fundamental issue is whether or not exploration and/or development is feasible under the conditions outlined in the fiscal system. The following pages outline the key aspects of contract negotiations and the numerous fiscal devices and systems designed to maximize host government profitability.

NEGOTIATIONS

The issue of the division of profits lies at the heart of contract/license negotiations. The purpose of fiscal structuring and taxation is to capture all the economic rent but also to provide a sufficient potential return for the oil companies. Unfortunately fiscal marksmanship is difficult. Structuring a fiscal system that will be appropriate or on-target under a variety of future (unknown) circumstances is nearly impossible.

Government Options

The objective of a host government is to maximize wealth from its natural resources by encouraging appropriate levels of exploration and development activity.

In order to accomplish this objective, governments must design fiscal systems that:
- Provide a fair return to the government and the industry
- Avoid undue speculation
- Limit undue administrative burden on government and industry
- Provide flexibility
- Create healthy competition and market efficiency

The design of an efficient fiscal system must take into consideration the political and geological risks as well as the potential rewards. One country may tax profits at a rate of 85% or more (like Indonesia or Malaysia), while another country may only have an effective tax rate of 40% (like the UK). Yet both countries may be efficiently extracting their resource rent regardless of the kind of system that is used. The real difference is in the level of profitability required on the part of the petroleum industry.

Malaysia is often touted as having one of the toughest fiscal systems in Southeast Asia. This is generally true but is balanced by the fact that Malaysia has good geological potential and robust GNP growth. The balance between prospectivity and fiscal terms is a fundamental theme in the industry. A lot of companies would love to explore in Malaysia and the government knows this. Governments are not the only ones who determine the difference between fair return and rent. The market works both ways.

The two primary economic aspects of contract/license negotiations are the work commitment and the fiscal terms. The work commitment represents hard *risk dollars*, while fiscal terms govern the allocation of revenues resulting from oil and gas production. Fiscal terms will also impact success ratios because fiscal terms have a strong bearing on development threshold field size, which is the difference between technical success and commercial success.

Work commitment	**Fiscal terms**
• Signature bonus	• Royalties
• Seismic acquisition	• Cost recovery (C/R)
• Drilling commitment	• Profit oil splits
	• Taxes
	• Government participation

Figure 1–2 shows how these elements influence the basic industry risk/reward relationship. This is a graphical representation of a simple two-outcome expected monetary value theory (EMV) model, also known as EV. The work commitment and signature bonus dominates the risk side of the equation, and the fiscal terms influence the success ratio estimate as well as the reward side of the equation.

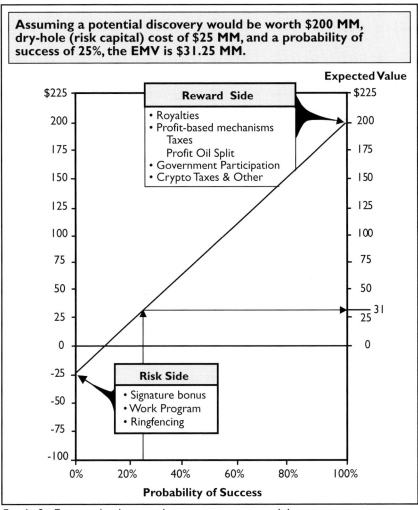

Assuming a potential discovery would be worth $200 MM, dry-hole (risk capital) cost of $25 MM, and a probability of success of 25%, the EMV is $31.25 MM.

Fig. 1–2 Expected value graph—two-outcome model

The reason that the fiscal terms influence the probability of drilling success is that the level of taxation will determine to a large extent how big a discovery must be to justify commercial development. As Figure 1–3 illustrates, the difference between technical and commercial success is the development threshold field size. If the probability of finding hydrocarbons is 20% but the accumulation must be greater than 25 MMBBLS to be economic, then the commercial success ratio will be substantially less than the technical success ratio. Commercial success ratios are always less. The royalty and tax rates as well as how they are structured can have a big impact on threshold field size.

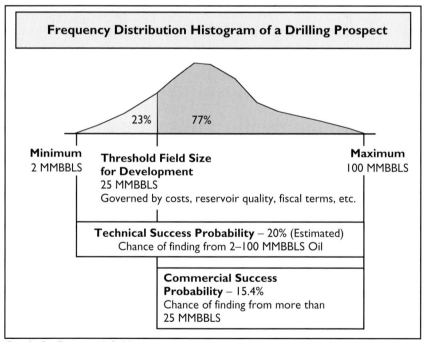

Fig. 1–3 Expected field size—oil

Figure 1–4 shows example *field development thresholds* from various countries and provinces. The predominant variable shown on this graph is the contractor share of profits or contractor take. The threshold field sizes range from as low as a couple million barrels to upwards of 100 MMBBLS in some of the more remote harsh environment regions.

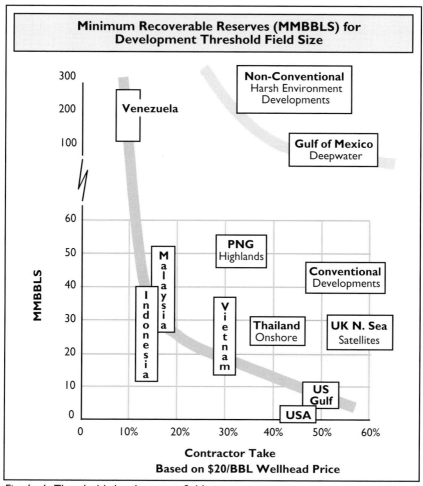

Fig. 1–4 Threshold development field size

While development thresholds may be on the order of 10 to 30 MMBBLS in some areas, exploration thresholds are ordinarily at least an order of magnitude greater. Explorationists must search for at least the 200–500 MMBBL fields.

Of the numerous kinds of contracts or fiscal arrangements in the world today, there are basic themes that fall under two main families: concessionary systems, and contractual systems. The taxonomy of petroleum fiscal systems is outlined in Figure 1–5. Determination of the appropriate system and then fiscal terms depends on the prospectivity of a block or region as shown in Figure 1–6. There is a balance between fiscal terms and prospectivity that must be considered.

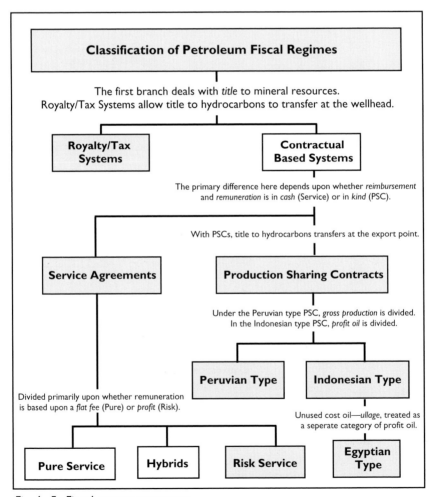

Fig. 1–5 Fiscal system taxonomy

There must be a balance between prospectivity and contract terms	
Balance Sheet	
Prospectivity	**Contract Terms**
• Field Size Distribution	• Type of System
• Success Probability	• Signature Bonus
• Petrophysical Characteristics	• Work Obligations
	• Duration and Relinquishment
• Fluid Properties Pressure Gradients	• Royalty
	• Government Take
• Aspect Ratios	• Effective Royalty Rate
• Water/Reservoir Depths	• Lifting Entitlement
• Exploration Drilling Costs	• Cost Recovery Limits
	• Ringfencing
• Transportation Costs	• Crypto Taxes
• Other Costs	• Allocation Strategy
• Country/Political Risks	• Other

Fig. 1–6 The balance sheet

The issue of ownership is the fundamental distinction between the concessionary and the contractual systems. Under a concessionary system, the oil company has title to the crude oil produced, against which it typically pays royalties and taxes. Under the contractual systems, which are ordinarily collectively referred to as PSCs, the government retains title to the mineral resources. This ownership issue drives not only the language and jargon of fiscal systems but the arithmetic as well. However, there are some PSCs that are identical to a concessionary system in all but the issue of ownership and the terminology used.

It is partially because of the concept of ownership of mineral resources that the term *contractor* has come into such wide use.

The earliest uses of the production-sharing concept occurred in the agriculture industry. Therefore, the term is used in the same context as *sharecropper* where ownership of the land and minerals is held by the government/landlord. The contractor or tenant/sharecropper is compensated out of production of minerals or grain, for example, according to a specific sharing arrangement. The term *contractor* therefore theoretically applies to PSCs or service agreements only, but with practical usage it cuts across the boundary between PSCs and concessionary systems. The term *concessionaire* in reference to the oil companies in a concessionary system might be technically correct, but it is not ordinarily used.

The ownership issue has one other element. Typically under contractual systems, once production equipment or facilities are landed in-country, commissioned, or placed-in-service, title to the equipment passes to the host government. However, this does not apply to leased equipment or service company equipment.

The contractual arrangements are divided into service contracts and PSCs. The difference between PSCs and risk service contracts depends upon whether or not the contractor receives compensation in cash or in kind (crude). This is a rather modest distinction and as a result, systems on this branch are commonly referred to as PSCs or PSAs.

For example, in the Philippines, the government alternately refers to their contractual arrangement as either a service contract or a PSC. The oil community does the same thing but more ordinarily refers to this system as a PSC. In a strict sense though, it would be more appropriate to classify it as a *risk service contract* because the contractor does not take title to hydrocarbons produced.

In a risk service contract the contractor gets a *share* of profits, not production, and therefore it is not a *production-sharing* contract. With that in mind, the term revenue-sharing or profit-sharing contract might be appropriate, but these terms are seldom used.

Variations on Two Themes

From a practical point of view, there are essentially the two basic families of systems—concessionary (royalty/tax [R/T]) and contractual (PSCs). Numerous variations and twists are found under both systems. The philosophical differences between the two systems have fostered a terminology unique to each. However, the terms are often simply different names for basic concepts.

Because of the modest differences between service agreements and PSCs, the study of PSCs effectively covers all aspects of contractual systems. The language and arithmetic of most PSCs and service agreements are basically identical.

Comparing and contrasting PSCs with concessionary systems, or more accurately *R/T systems*, provides an excellent foundation for understanding the bulk of the subject of petroleum fiscal system economics.

Trends in Fiscal System Development

There was an obvious trend in the 1980s and 1990s. Most countries developing petroleum fiscal systems were opting for the PSC. Certainly the bottom line could be the same with an R/T system, depending upon the aggregate level of royalties, taxes, and levies. The philosophical and political aspects come into play, though, and the advantage is toward the PSC.

Progressive Systems

Systems with flexible terms are becoming standard. There are many advantages for both the host government and the contractor with systems that attempt to encompass a range of economic conditions, i.e. both highly profitable and marginal fields.

The most common method used for creating a flexible fiscal system is with sliding scale terms. There are nearly as many kinds of sliding scales as there are contracts. Most sliding scale systems impose a progressively smaller share of profit oil (P/O) for the contractor as production rates increase. This theoretically allows equitable terms for development of both large and smaller fields. Contracts may subject a number of terms to sliding scales, which may be determined by one or more conditions.

Some contracts will tie more than one variable to a sliding scale such as: C/R limits, P/O splits, and royalties. Table 1–1 shows the diversity of contract elements that are subject to sliding scales and the factors that will trigger a change.

Table 1–1 Flexible Contract Terms and Conditions

Flexible Contract Terms and Conditions	
Contract Terms Subject to Sliding Scales	**Factors and Conditions That Trigger Sliding Scales**

• Profit Oil Split
• Royalty
• Bonuses
   Cost Recovery Limits
   Tax Rates
• Most Common
	• Production Rates
	• Water Depth
	• Cumulative Production
	• Oil Prices
	   Age or Depth of Reservoirs
	   Onshore vs. Offshore
	   Remote Locations
	• Oil vs. Gas
	   Crude Quality (Gravity)
	• Rate of Return

The usual approach is an incremental sliding scale based upon average daily production. The following example shows a sliding scale royalty that steps up from 5% to 15% on 10,000 BOPD tranches of production. If average daily production gets to 15,000 BOPD, the aggregate effective royalty paid by the contractor would be 6.667% (10,000 BOPD at 5% + 5,000 BOPD at 10%).

Sliding Scale Royalty

Average Daily Production	Royalty
Up to 10,000 BOPD	5%
10,001 to 20,000 BOPD	10%
Above 20,000 BOPD	15%

Sometimes misconceptions arise when it is assumed that in a case like this, once production exceeds 10,000 BOPD, all production would be subject to the 10% royalty. No. These sliding scales do not work that way.

The opposite of a pure bonus bid approach would be pure profits-based taxation. This is more practical. Of the four main means by which governments capture rent (*see* Fig. 1–7), the distribution is roughly as follows:

Percentage of countries that use:

Signature Bonuses 40%

Royalties 75% 1/3 of them are *sliding*
Of the flat ones, 2/3 are less than 12.5%
World average is 7%

Government Participation 50% Average working interest participation is 30% for those countries with the participation option

Profits-based Mechanisms

Taxes 90% 75% are direct
25% are *deemed paid* or *in lieu*
1/3 have additional taxes

P/O Splits 50% 85% of these are *sliding*

Other 25% have withholding taxes
7% have DMOs

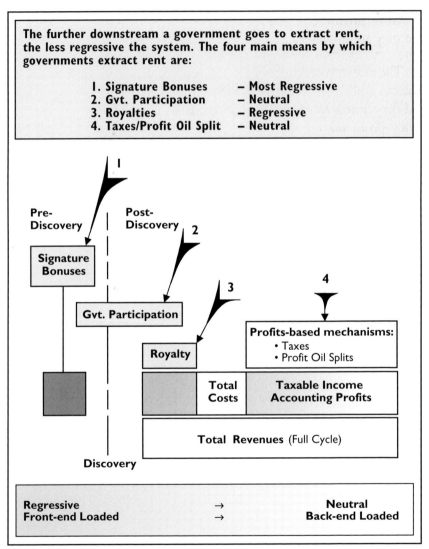

Fig. 1–7 Tax base spectrum

Governments base most of their take on profits-based mechanisms, and they are moving even further in this direction.

The ultimate objective of a flexible system is to create a framework that can honor the mutuality of interest between the host government and the contractor and provide an equitable arrangement for both the highly profitable and the less profitable discoveries. A stiff, inflexible royalty is the antithesis of flexibility.

The acid test for the flexibility and fairness of any fiscal system is whether or not an equitable, profitable arrangement can be achieved for both the host government and the contractor under a variety of conditions. Unfortunately project profitability is too often a function of government take, and this usually hurts both the government and the oil companies. The better arrangements are where government take is a function of profitability. Figure 1–8 illustrates this aspect of flexibility showing how government take increases as project profitability increases. This is the objective of sliding scales as well as rate-of-return (ROR) systems. Inflexible systems with high royalty rates can work in just the opposite way. The sliding scale examples listed previously are attempts at creating a progressive taxation system based upon some proxy of profitability. They are found under both concessionary and contractual systems. Many of the features indirectly address project profitability but true ROR triggers address virtually all aspects of profitability. ROR contracts or ROR systems are rare but are distinctive enough to warrant special attention.

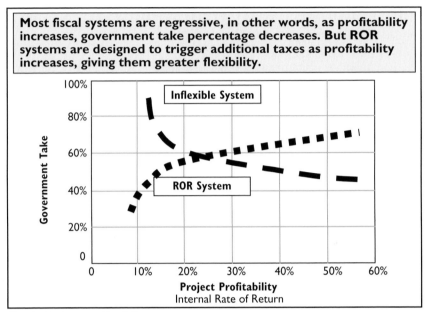

Fig. 1–8 System flexibility

R/T Systems

The term *concession* has a lot of negative connotations these days. *R/T system* is becoming the preferred terminology for those who care to be politically correct.

The term is quite appropriate as a descriptive term because most R/T systems truly are not much more than just that—a combination of royalty and taxes. Sliding scale features abound, and there are numerous features that are peculiar to one country or another, but even the most complicated R/T systems are usually fairly simple.

R/T System Flow Diagram

Figure 1–9 depicts the typical revenue distribution under an R/T system. The diagram illustrates the hierarchy of royalties, deductions, and in this example, two layers of taxation. For illustration purposes, a single barrel of oil is forced through the system.

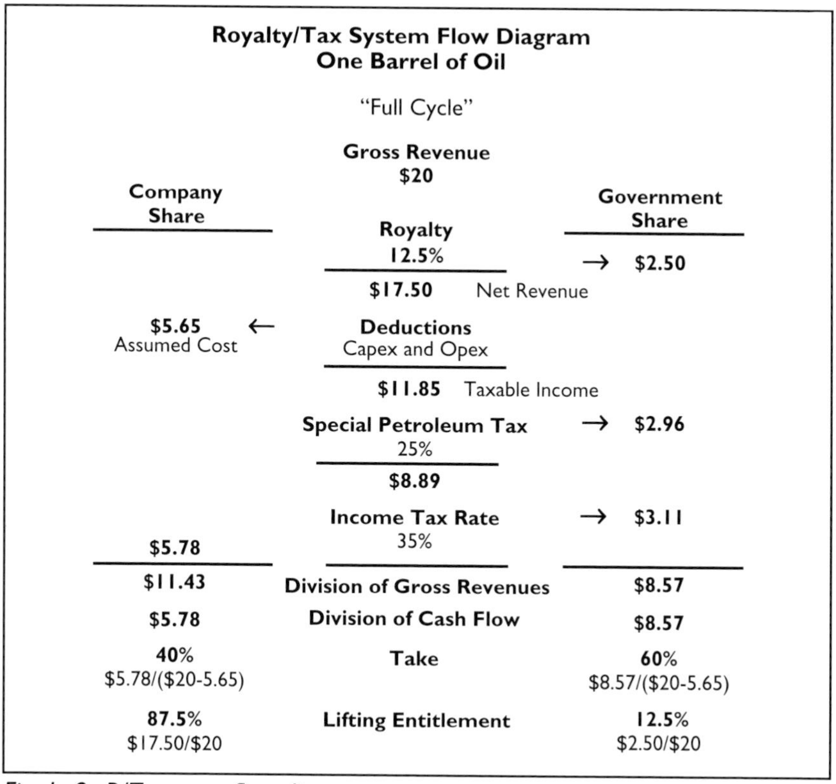

Royalty/Tax System Flow Diagram
One Barrel of Oil

"Full Cycle"

Gross Revenue
$20

Company Share		Government Share
	Royalty	
	12.5%	→ $2.50
	$17.50 Net Revenue	
$5.65 ←	**Deductions**	
Assumed Cost	Capex and Opex	
	$11.85 Taxable Income	
	Special Petroleum Tax	→ $2.96
	25%	
	$8.89	
	Income Tax Rate	→ $3.11
$5.78	35%	
$11.43	**Division of Gross Revenues**	$8.57
$5.78	**Division of Cash Flow**	$8.57
40%	**Take**	60%
$5.78/($20-5.65)		$8.57/($20-5.65)
87.5%	**Lifting Entitlement**	12.5%
$17.50/$20		$2.50/$20

Fig. 1–9 R/T system flow diagram

First—Royalty

The royalties come right off the top. Royalties are a well understood concept, and while there are some rather exotic variations on the royalty theme, they are rare. Both R/T systems and PSCs will sometimes allow a netback on transportation costs. In this example, a 12.5% royalty is used.

Second—Deductions

Before calculation of taxes, the contractor is allowed to deduct operating costs, depreciation, depletion & amortization (DD&A) and intangible drilling costs (IDCs) from net revenues. Usually only depreciation is used for tangible capital costs (as is the case in the economic models in this book) but the term *DD&A* usually refers collectively to the capitalization of any costs.

Third—Taxation

Revenues remaining after royalty and deductions are called *taxable income*, and in this example, there are a couple of layers of taxation, namely: a 25% supplementary petroleum tax (SPT) and 35% corporate income tax (CIT). SPT is deductible against income taxes, so the effective tax rate is 51.2%. Depreciation, for tax calculation purposes is a 5-year straight line decline (SLD) starting in the year production begins. Notice for each of these taxes there is a column for carrying forward (C/F) tax losses—called a tax loss carry forward (TLCF).

With tax deductions, the contractor share of gross revenues is 57% ($11.43/$20). The contractor share of profits is 40% ($5.78/$14.35). This is what is called *contractor take*. The calculation of take focuses on the division of profits.

Table 1–2 summarizes the respective takes from the cash flow projection. Government take is 60%. A quick-look estimate from the flow diagram yields almost exactly the same result—as well it should. The cash flow assumptions for a 100 MMBBL oilfield are summarized in Table 1–3. The only difference between the flow diagram and the detailed cash flow projection in regard to this estimate of take is that the bonus ($5 MM) was ignored in the flow diagram. Had there been a significant difference between the take calculations between the cash flow model shown in Tables 1–4 and 1–5 and the flow diagram, then this would indicate a possible problem in either the model or the flow diagram. Notice in Table 1–4 that the contractor cash flow of $574,438 M yields a discounted cash flow (DCF) value of only $142,492 M. Comparing the division of government and contractor share of DCF will yield a lower contractor take. This is always the case—that government take goes up when factoring-in time value of money.

Table 1–2 R/T System Cash Flow Model Summary & Analysis

The R/T system summary simply deducts costs from gross revenue to get profit, deducts the components of government take, (bonus, royalty, and 2 layers of tax) totaling $860,563. Contractor profit is $574,438. The take calculation yields a 60/40% split in favor of the Government, i.e. Government take is 60%.

Gross Revenues	$2,000,000	
Total Costs	-565,000	(28.25%)
Total Profit	$1,435,000	
Bonus	- 5,000	
Royalties (12.5%)	- 250,000	
SPT Tax (25%)	- 296,250	
Income Tax (35%)	- 309,314	860,563 (Total Gvt. Take)
Company Cash Flow	$574,438	
Company Take	40%	($574,438/1,435,000)
Government Take	60%	(100%—Company Take)
		Or ($860,563/1,435,000)
		(Undiscounted)
Government DCF (12.5%)	$308,854	
Company DCF (12.5%)	$142,492	
Government Take		
Discounted (12.5%)	68.4%	$308,854/(308,854+142,492)

SPT = supplementary petroleum tax   DCF = discounted cash flow

Table 1–3 Oilfield X Vital Statistics

The Basic Unit of Production
The anticipated field size distribution in any given province is an important part of fiscal system design. The following parameters are used throughout this book for the economic models depicting various fiscal arrangements and devices.

Recoverable Reserves	**100 MMBBLS**	
Peak Production Rate	**34,000 BOPD**	(12,400 MBBLS—Year 6)
Production/Reserves ratio	12.4%	
Decline Rate	12.5%	
Field Life	17 Years	
Oil Price	$20/BBL	
Capital Costs (Capex)	$300 MM	
Capex/BBL	$3.00/BBL	
Operating Costs (Opex)	$265 MM	$4 MM/Year Fixed
		$2/BBL Variable
Opex/BBL	$2.65/BBL	Average (full cycle)

Table 1–4 Sample Royalty/Tax System Cash
Flow Projection Field X Development Feasibility Study

Year	Annual Oil Production (MBBLS)	Oil Price ($/BBL)	Gross Revenues ($M)	Royalty 12.5% ($M)	Net Revenue ($M)	Capital Costs ($M)	Operating costs ($M)	Depreciation ($M)	SPT TLCF ($M)	SPT Deductions ($M)
	A	B	C	D	E	F	G	H	I	J
1	0	$20				30,000			0	0
2	0	$20				40,000			0	0
3	578	$20	11,560	1,445	10,115	100,000	3,156	34,000	0	10,115
4	6,100	$20	122,000	15,250	106,750	60,000	16,200	46,000	27,041	89,241
5	9,420	$20	188,400	23,550	164,850	70,000	22,840	60,000		82,840
6	12,400	$20	248,000	31,000	217,000		28,800	60,000		88,800
7	10,850	$20	217,000	27,125	189,875		25,700	60,000		85,700
8	9,494	$20	189,880	23,735	166,145		22,988	26,000		48,988
9	8,307	$20	166,140	20,768	145,373		20,614	14,000		34,614
10	7,269	$20	145,380	18,173	127,208		18,538			18,538
11	6,360	$20	127,200	15,900	111,300		16,720			16,720
12	5,565	$20	111,300	13,913	97,388		15,130			15,130
13	4,869	$20	97,380	12,173	85,208		13,738			13,738
14	4,261	$20	85,220	10,653	74,568		12,522			12,522
15	3,728	$20	74,560	9,320	65,240		11,456			11,456
16	3,262	$20	65,240	8,155	57,085		10,524			10,524
17	2,854	$20	57,080	7,135	49,945		9,708			9,708
18	2,498	$20	49,960	6,245	43,715		8,996			8,996
19	2,185	$20	43,700	5,463	38,238		7,370			7,370
20										
Total	100,000		2,000,000	250,000	1,750,000	300,000	265,000	300,000		565,000

Year	SPT Base ($M)	SPT Tax 25% ($M)	Bonus ($M)	Income Tax TLCF ($M)	Taxable Income ($M)	Income Tax 35% ($M)	Contractor Cash Flow ($M)	
							Undiscounted	12.5% DCF
	K	L	M	N	P	Q	R	S
1	0	0	5,000	0	(5,000)	0	(35,000)	(32,998)
2	0	0		(5,000)	(5,000)	0	(40,000)	(33,522)
3	0	0		(5,000)	(32,041)	0	(93,041)	(69,310)
4	17,509	4,377		(32,041)	8,132	2,846	23,327	15,446
5	82,010	20,503			61,508	21,528	29,980	17,646
6	128,200	32,050			96,150	33,653	122,498	64,090
7	104,175	26,044			78,131	27,346	110,785	51,522
8	117,157	29,289			87,868	30,754	83,114	34,358
9	110,759	27,690			83,069	29,074	67,995	24,985
10	108,670	27,167			81,502	28,526	52,976	17,303
11	94,580	23,645			70,935	24,827	46,108	13,387
12	82,258	20,564			61,693	21,593	40,101	10,349
13	71,470	17,867			53,602	18,761	34,841	7,993
14	62,046	15,511			46,534	16,287	30,247	6,168
15	53,784	13,446			40,338	14,118	26,220	4,752
16	46,561	11,640			34,921	12,222	22,698	3,657
17	40,237	10,059			30,178	10,562	19,616	2,809
18	34,719	8,680			26,039	9,114	16,926	2,155
19	30,868	7,717			23,151	8,103	15,048	1,703
20								
Total	1,185,003	296,250				309,313	574,438	142,492

TLCF = tax loss carry forward
SPT = supplementary petroleum tax
DCF = discounted cash flow

Table 1–5 Sample Royalty/Tax System Cash Flow Projection
Government Cash Flow

Year	Bonuses ($M)	Royalty ($M)	SPT Tax 25% ($M)	Income Tax 35% ($M)	Government Cash Flow ($M) Undiscounted	12.5% DCF
	M	D	L	Q		
1	5,000		0	0	5,000	4,714
2			0	0	0	0
3		1,445	0	0	1,445	1,076
4		15,250	4,377	2,846	22,473	14,881
5		23,550	20,503	21,528	65,580	38,600
6		31,000	32,050	33,653	96,703	50,594
7		27,125	26,044	27,346	80,515	37,444
8		23,735	29,289	30,754	83,778	34,633
9		20,768	27,690	29,074	77,531	28,489
10		18,173	27,167	28,526	73,866	24,126
11		15,900	23,645	24,827	64,372	18,689
12		13,913	20,564	21,593	56,069	14,470
13		12,173	17,867	18,761	48,801	11,195
14		10,653	15,511	16,287	42,451	8,656
15		9,320	13,446	14,118	36,884	6,685
16		8,155	11,640	12,222	32,018	5,159
17		7,135	10,059	10,562	27,756	3,975
18		6,245	8,680	9,114	24,038	3,060
19		5,463	7,717	8,103	21,282	2,408
20						
Total	5,000	250,000	296,250	309,314	860,562	308,854

A) Production Profile Thousands (M) barrels/year	**J) SPT Deductions** = (G + H + I) up to 100% of E
B) Crude Price	**K) SPT Base** = (C – D – J)
C) Gross Revenues Thousands of dollars ($M)	**L) SPT 25%** = (K * .25)
D) Royalty 12.5% = (C * .125)	**M) Signature Bonus**
E) Net Revenues = (C – D)	**N) CIT Loss Carried Forward** (See Column P)
F) Capital Costs	**P) Taxable Income** = (C – D – G – H – L – N)
G) Operating Costs (Expensed)	**Q) Income Tax (35%)** = (if P > 0, P * .35, otherwise 0)
H) Depreciation of Capital Costs (5-year SLD)	**R) Company Cash Flow** = (E – F – G – L – M – Q)
I) SPT Loss Carried Forward (if G + H + I > E)	**T) Government Cash Flow** = (D + L + M + Q)
TLCF = tax loss carry forward	**CIT** = corporate income tax
SPT = supplementary petroleum tax	**DCF** = discounted cash flow
SLD = straight line decline	

Another dimension of contractor take comes from the effect of royalties or any taxes that are levied on gross revenues and not profits. With different levels of profitability, fiscal systems with royalties can yield different government/contractor takes. An example summary of the R/T system structure in a single accounting period (Year 5) is summarized in Table 1–6. The contractor take calculations are shown in Table 1–7. Also the basic equations for R/T systems are summarized in Table 1–8.

Table 1–6 R/T System Structure

Single Accounting Period—Year 5

In any given accounting period (if there is depreciation), net income will not be equal to cash flow. In accounting periods where there is no depreciation, net income will equal cash flow. And typically, full-cycle cumulative cash flow will equal net income.

Terminology	Gross	$/BBL	Operations
Revenues	$188,400	$20.00	
	-23,550	-2.50	12.5% (1/8th) Royalty
Net Revenue	164,850	17.50	
Before-Tax	-22,840	-2.42	Operating Costs
Operating Income	142,010	15.08	(Includes Abandonment)
Before-tax	-60,000	-6.37	Depreciation
Net Income	82,010	8.71	
	-20,503	- 2.18	25% Special Petroleum Tax
	61,508	6.53	
After-Tax	-21,528	- 2.29	35% Corporate Income Tax
Net Income	39,980	$4.24	
	+60,000	+6.37	Depreciation
After-Tax	-70,000	-7.43	Tangible Capital Costs
Cash Flow	$29,980	$3.18	

Table 1–7 Different Perspectives on the Example R/T System

This example compares three cases with costs varying from high to low, and profits varying from low to high. The resultant Government Take, drops as profits increase, demonstrating the regressive effect of the royalty on this R/T system.

Gross Revenues = 100% Full Cycle
Royalty = 12.5%
Taxes = 25% SPT + 35% CIT = Effective Tax Rate = 51.25%

Costs as a percentage of Gross Revenues—Three Scenarios:

60% - High-cost case
↓
   30% - Low-cost case
   ↓
      0% - At the margin
      ↓

High-Cost Case	Low-Cost Case	Zero-Cost Margin	
A	B	C	← [See Fig. 1–10]
100.00%	100.00%	100.00%	Gross Revenues
-12.5	- 12.5	- 12.5	Royalty
87.5	87.5	87.5	Net Revenues
-60.00	-30.00	0.00	Total Costs (Deductions)
27.5	57.5	87.5	Taxable Income
-14.09	-29.46	-44.84	Total Taxes 51.25%
13.41%	28.04%	42.66%	Contractor Cash Flow
33.5%	40.0%	42.7%	Contractor Take
			[Contractor Cash Flow ÷ (Gross Revenues – Costs)]
66.5%	60.0%	57.3%	Government Take

Table 1–8 Basic Equations—R/T Systems

Gross Revenues	=	Total Oil and Gas Revenues
Net Revenues	=	Gross Revenues – Royalties
Net Revenue (%)	=	100%—Royalty Rate (%)
Taxable Income	=	Gross Revenues
		- Royalties
		- Operating Costs (OPEX)
		- Intangible Capital Costs *
		- Depreciation, Depletion, & Amortization (DD&A)
		- Investment Credits (if allowed)
		- Interest on Financing (if allowed)
		- Tax Loss Carry Forward (TLCF)
		- Abandonment Cost Provision (included in Opex)
		- Bonuses **
Net Cash Flow (after-tax)	=	Gross Revenues
		- Royalties
		- Tangible Capital Costs
		- Intangible Capital Costs*
		- Operating Costs
		- Bonuses
		- Taxes

* In many systems, no distinction is made between *operating costs* and *intangible capital costs* and both are *expensed*.

** Bonuses are not always deductible for tax calculation purposes.

The take calculations in Table 1–7 illustrate the impact of the 12.5% royalty. This is a regressive fiscal structure. The lower the profitability, the higher the effective tax rate. This is because of the royalty. It is based on gross revenues. The step-by-step allocation of revenues under high, low, and zero-cost cases is shown here with government take decreasing with increased profitability. A graphical representation of the regressive nature of the royalty is shown in Figure 1–10.

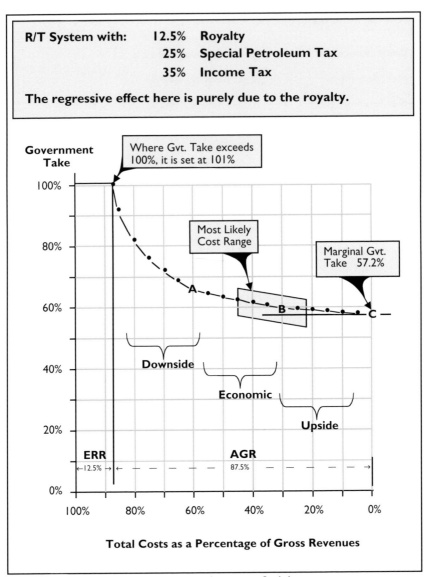

Fig. 1–10 *R/T System government take vs. profitability*

Contractor Take—The Common Denominator

Contractor take or government take statistics give a quick measure of comparison between one fiscal system and another. They focus exclusively on the division of profits and are common quick-look techniques and comparative tools. They also correlate directly with reserve values, field size thresholds and other measures of relative economics.

The main limitation of the contractor take statistic is that it does not account for other aspects of a given fiscal system such as C/R limits, ringfencing (where each license area is treated as a separate cost pool for tax and C/R calculations) investment credits, work programs, bonuses, DMOs, etc. If viewed from the government perspective, it is referred to as government take. The *complement of government take* (meaning 1 minus government take) is contractor take. For example, if government take is 75%, then contractor take is 25% (1-0.75).

The concept of government or contractor take is slightly abstract and can be misleading because it deals only with the division of profit. For example, in a system with an 80/20 split in favor of the government (or government take of 80%), the contractor may very well end up with more than 50% of the crude oil because of C/R. Some crude goes to the contractor as reimbursement of costs incurred (C/R) and some as a share of profit.

Even under a system like the Indonesian PSC with its 85/15 split in favor of the government, the contractor will end up with a 40–50% share of production. However, the Indonesian 15% contractor take is a measure of the share of profits and that is a more meaningful number.

An example is shown in Figure 1–11 that depicts the relationship between producing-reserve values and contractor take. Reserve values correlate quite closely. Limits on C/R and other factors also affect reserve values but not nearly as much as this factor.

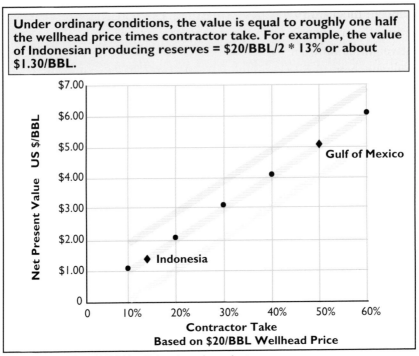

Under ordinary conditions, the value is equal to roughly one half the wellhead price times contractor take. For example, the value of Indonesian producing reserves = $20/BBL/2 * 13% or about $1.30/BBL.

Fig. 1–11 Value ($/BBL) of developed producing reserves

PSCs

It was probably inevitable that PSCs would become an important part of the international petroleum industry. The concept dates back so far that it has distilled itself into the collective human consciousness. It hails from the agricultural world.

At first, PSCs appear to be quite different than R/T systems. PSCs have major symbolic and philosophical differences, and they appear to fill some psychological niche, but they serve more of a political function than anything else. The terminology is certainly distinct, but these systems are not really that different from a financial/mechanical point of view. In fact, as far as the mechanics and arithmetic are concerned, the similarities are dramatic and by far outweigh the differences. For all practical purposes, there is only one small mechanical difference—the C/R limit. Furthermore, from a legal point of view, the differences are not dramatic.

The arithmetic, economic, and financial aspects of a simple PSC are evaluated first. Most non-financial features of a PSC are similar to those found under other systems. Therefore, these common elements are discussed in detail later.

Of the numerous production-sharing arrangements there are common elements. The essential characteristic of course is that of state ownership of the resources. The contractor receives a share of production for services performed.

Now, as more countries open their doors to the petroleum industry, they use PSCs as opposed to concessionary systems.

The first PSC was signed by IIAPCO in August 1966, with Permina, the Indonesian National Oil Company at that time (now Pertamina).

This is when oil companies became *contractors*. This contract embodied the basic features of the production-sharing concept:

- Title to the hydrocarbons remained with the state.
- Permina maintained management control, and the contractor was responsible to Permina for execution of petroleum operations in accordance with the terms of the contract.
- The contractor was required to submit annual work programs and budgets for scrutiny and approval by Permina.
- The contract was based on production sharing and not a profit-sharing basis.
- The contractor provided all financing and technology required for the operations and bore the risks.

- During the term of the contract, after allowance for up to a maximum of 40% of annual oil production for recovery of costs, the remaining production was shared 65/35% in favor of Permina. The contractors' taxes were paid out of Permina's share of P/O.

- All equipment purchased and imported into Indonesia by the contractor became the property of Permina. Service company equipment and leased equipment was exempt.

These features continue to outline the nature of the government/contractor relationships under PSCs or service agreements. It is a formula that is popular with many governments.

An example PSC is outlined in Table 1–9 with a summary of economic and financial analysis.

Table 1–9 Example PSC

The primary components of this simple PSC include the bonus, royalty, cost recovery limit, profit oil split and taxes.		
Summary of Commercial Terms		
Signature Bonus	$5 MM	
Tax deductible but not cost recoverable		
Royalty Rate	10%	
Cost Recovery Limit	50%	
Government Share Profit Oil	60%	
Corporate Income Tax (CIT)	30%	
Depreciation Rate	5 year straight line (20%/year)	
Analysis Summary		
	Downside	**Economic**	**Upside**
Government Take (Undiscounted)	90%	76%	75%
Government Take @ 12.5% Discounted Cash Flow (DCF)		86.5%	
Marginal Government Take		74.8%	
Effective Royalty Rate (ERR)		34%	
Access to Gross Revenues (AGR)		66%	
Entitlement Index		53%	
Savings Index		28¢	

PSC Flow Diagrams

Figure 1–12 shows a flow diagram of an example PSC. It illustrates the terminology and hierarchy of arithmetic that would be experienced in a given accounting period but represents average full-cycle revenues and costs. For illustration, one barrel of oil is followed through the system. An average (full-cycle) cost (capital and operating) of $5.65/BBL is assumed here as before.

PSC Flow Diagram
One Barrel of Oil

"Full Cycle"

	Gross Revenues **$20**	
Contractor **Share**		**Government** **Share**
	Royalty 10%	→ **$2.00**
	$18.00	
$5.65 ← Assumed Costs	**Cost Recovery** 50% Limit	
	$12.35 **Profit Oil**	
$4.94 ←	**Profit Oil Split** 40/60%	→ **$7.41**
($1.48) →	**Tax Rate** 30%	→ **$1.48**
$3.46		
$9.11	Division of Gross Revenues	**$10.89**
$3.46	Division of Cash Flow	**$10.89**
24% $3.46/($20-5.65)	Take	76% $10.89/($20-5.65)
53% ($5.65+4.94)/$20	Entitlement	47% ($2.00+7.41)/$20

Fig. 1–12 PSC flow diagram

Bonus

In this example, a $5 MM bonus is due and payable upon signature. In the grand scheme, it may not amount to much in the context of a

100-MMBBL field. It comes to only 5¢/BBL. Furthermore, with an oil price of $20/BBL and $2 billion in revenues it amounts to only one quarter of 1%. Therefore it is easily ignored in the back-of-the-envelope calculations to follow; but in the absence of a discovery, it takes on added significance.

First—Royalty

The royalty comes right off the top just as it would in an R/T system. This example uses a 10% royalty. Royalties are not extremely common in PSCs but many have them.

Second—C/R

Before sharing of production, the contractor is allowed to *recover costs* out of net revenues. However, most PSCs will place a limit on how much production (or revenues) will be made available for the recovery of costs in any given accounting period. This is known as the *C/R limit*. For example, in the flow diagram, C/R is limited to 50% of gross revenues. If operating costs and depreciation amount to more than that, the balance is carried forward and recovered later. It means that there is a limit to the amount of deductions that can be taken in any given accounting period. Most PSCs allow virtually unlimited carry forward (C/F). From a mechanical point of view, the C/R limit is the only true distinction between R/T systems and PSCs.

Third—P/O Split

Revenues remaining after royalty and C/R are referred to as P/O or *profit gas*. The analog in a concessionary system would be taxable income. The terminology is precise because of the ownership issue. The term taxable income implies ownership that does not exist yet under a PSC. The contractor has nothing to tax—not yet.

In this example, the contractor's *share* of P/O is 40%. If this were a service agreement with the contractors' share of revenues equal to 40%, it would likely be called the *service fee*—not P/O. The government 60% share has all of the characteristics of an accounting-profits-based tax.

Fourth—Taxes

The tax rate of 30% in this flow diagram appears to apply to the P/O. It is acceptable to do this when thinking in terms of *full-cycle* economics. On the average over the life of a field, the accounting profits subject to ordinary taxes will be equal to the company share of P/O. However, the P/O ordinarily does not constitute the tax base. In any given accounting period, a company will receive a share of P/O if there is a C/R limit but the company may not necessarily be in a tax-paying position. This is important when considering the royalty effect of the C/R limit (*see* Fig. 1–13) in conjunction with the P/O split. Table 1–9 summarizes basic statistics of the example PSC described above and a cash flow model is provided in Tables 1–10 and 1–11.

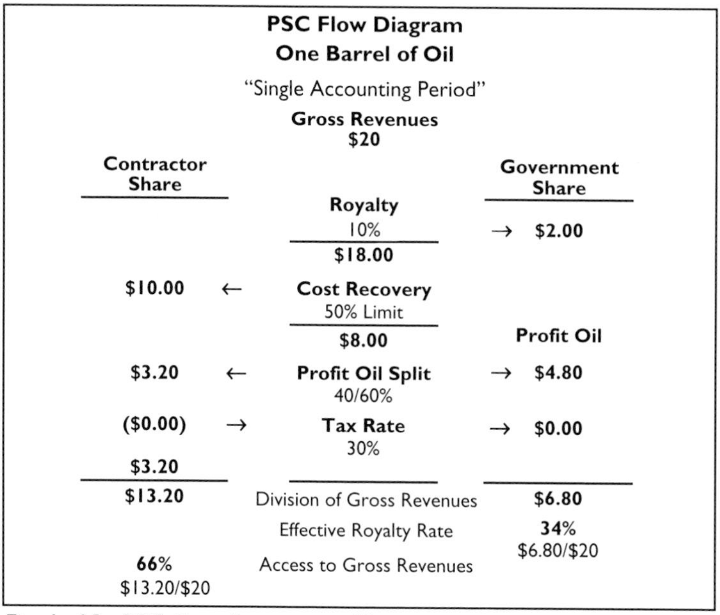

PSC Flow Diagram
One Barrel of Oil

"Single Accounting Period"

Contractor Share		Gross Revenues $20	Government Share	
		Royalty 10% $18.00	→	$2.00
$10.00	←	Cost Recovery 50% Limit $8.00	Profit Oil	
$3.20	←	Profit Oil Split 40/60%	→	$4.80
($0.00)	→	Tax Rate 30%	→	$0.00
$3.20				
$13.20		Division of Gross Revenues	$6.80	
		Effective Royalty Rate	34% $6.80/$20	
66% $13.20/$20		Access to Gross Revenues		

Fig. 1–13 PSC flow diagram—effective royalty rate calculation

Table 1–10 Sample Production-Sharing Contract Cash Flow Projection
Field X Development Feasibility Study

Year	Annual Oil Production (MBBLS)	Oil Price ($/BBL)	Gross Revenues ($M)	Royalty 10% ($M)	Net Revenue ($M)	Capital Costs ($M)	Op. Costs ($M)	Depreciation ($M)	C/R C/F ($M)	C/R ($M)
	A	B	C	D	E	F	G	H	I	J
1	0	$20				30,000				0
2	0	$20				40,000				0
3	578	$20	11,560	1,156	10,404	100,000	3,156	34,000		5,780
4	6,100	$20	122,000	12,200	109,800	60,000	16,200	46,000	31,376	61,000
5	9,420	$20	188,400	18,840	169,560	70,000	22,840	60,000	32,576	94,200
6	12,400	$20	248,000	24,800	223,200		28,800	60,000	21,216	110,016
7	10,850	$20	217,000	21,700	195,300		25,700	60,000		85,700
8	9,494	$20	189,880	18,988	170,892		22,988	26,000		48,988
9	8,307	$20	166,140	16,614	149,526		20,614	14,000		34,614
10	7,269	$20	145,380	14,538	130,842		18,538			18,538
11	6,360	$20	127,200	12,720	114,480		16,720			16,720
12	5,565	$20	111,300	11,130	100,170		15,130			15,130
13	4,869	$20	97,380	9,738	87,642		13,738			13,738
14	4,261	$20	85,220	8,522	76,698		12,522			12,522
15	3,728	$20	74,560	7,456	67,104		11,456			11,456
16	3,262	$20	65,240	6,524	58,716		10,524			10,524
17	2,854	$20	57,080	5,708	51,372		9,708			9,708
18	2,498	$20	49,960	4,996	44,964		8,996			8,996
19	2,185	$20	43,700	4,370	39,330		7,370			7,370
20										
Total	100,000		2,000,000	200,000	1,800,000	300,000	265,000	300,000		565,000

Year	Total Profit Oil ($M)	Gvt Share ($M)	Company Share ($M)	Bonus ($M)	TLCF ($M)	Taxable Income ($M)	Income Tax 30% ($M)	Contractor Cash Flow Undiscounted	12.5% DCF
	K	L	M	N	O	P	Q	R	S
1	0	0		5,000		(5,000)		(35,000)	(32,998)
2	0	0			(5,000)	(5,000)		(40,000)	(33,522)
3	4,624	2,774	1,850		(5,000)	(34,526)		(95,526)	(71,161)
4	48,800	29,280	19,520		(34,526)	(16,206)		4,320	2,861
5	75,360	45,216	30,144		(16,206)	25,298	7,589	23,915	14,076
6	113,184	67,910	45,274			66,490	19,947	106,543	55,742
7	109,600	65,760	43,840			43,840	13,152	90,688	42,175
8	121,904	73,142	48,762			48,762	14,628	60,133	24,858
9	114,912	68,947	45,965			45,965	13,789	46,175	16,967
10	112,304	67,382	44,922			44,922	13,476	31,445	10,271
11	97,760	58,656	39,104			39,104	11,731	27,373	7,947
12	85,040	51,024	34,016			34,016	10,205	23,811	6,145
13	73,904	44,342	29,562			29,562	8,868	20,693	4,747
14	64,176	38,506	25,670			25,670	7,701	17,969	3,664
15	55,648	33,389	22,259			22,259	6,678	15,581	2,824
16	48,192	28,915	19,277			19,277	5,783	13,494	2,174
17	41,664	24,998	16,666			16,666	5,000	11,666	1,671
18	35,968	21,581	14,387			14,387	4,316	10,071	1,282
19	31,960	19,176	12,784			12,784	3,835	8,949	1,013
20									
Total	1,235,000	741,000	494,000				146,700	342,300	60,736

C/R = cost recovery DCF = discounted cash flow
C/F = carry forward TLCF = tax loss carry forward

Table 1–11 Sample Production-Sharing Contract Cash Flow Projection
            Government Cash Flow

Year	Bonuses ($M)	Royalty 10% ($M)	Gvt. 60% Profit Oil ($M)	Income Tax 30% ($M)	Government Cash Flow ($M)	
					Undiscounted	12.5% DCF
	N	D	L	Q	T	U
1	5,000				5,000	4,714
2					0	0
3		1,156	2,774		3,930	2,928
4		12,200	29,280		41,480	27,467
5		18,840	45,216	7,589	71,645	42,170
6		24,800	67,910	19,947	112,657	58,941
7		21,700	65,760	13,152	100,612	46,791
8		18,988	73,142	14,628	106,759	44,133
9		16,614	68,947	13,789	99,351	36,507
10		14,538	67,382	13,476	95,397	31,159
11		12,720	58,656	11,731	83,107	24,129
12		11,130	51,024	10,205	72,359	18,674
13		9,738	44,342	8,868	62,949	14,440
14		8,522	38,506	7,701	54,729	11,160
15		7,456	33,389	6,678	47,523	8,614
16		6,524	28,915	5,783	41,222	6,642
17		5,708	24,998	5,000	35,706	5,114
18		4,996	21,581	4,316	30,893	3,933
19		4,370	19,176	3,835	27,381	3,098
20						
Total	5,000	200,000	741,000	146,700	1,092,700	390,612

A) **Production Profile** Thousands (M) barrels/year
B) **Crude Price**
C) **Gross Revenues** Thousands of dollars ($M)
D) **Royalty 10%** = (C * .10)
E) **Net Revenues** = (C – D)
F) **Capital Costs**
G) **Operating Costs** (Expensed)
H) **Depreciation** of Capital Costs (5-year SLD)
I) **C/R C/F** (if G + H + I > 50% of C)
J) **Cost Recovery** = (G + H + I) up to 50% of C

K) **Total Profit Oil** = (C – D – J)
L) **Government Share P/O 60%** = (K * .60)
M) **Contractor Share P/O 40%** = (K – L)
N) **Signature Bonus**
O) **TLCF** (See Column P)
P) **Taxable Income** = (C – D – G – H – L – N – O)
Q) **Income Tax (30%)** = [if P > 0, P * .30]
R) **Company Cash Flow** = (E – F – G – L – N – Q)
T) **Government Cash Flow** = (D + L + N + Q)

C/R = cost recovery   P/O = profit oil   DCF = discounted cash flow
C/F = carry forward   TLCF = tax loss carry forward

Government Take

With C/R, the contractor's gross share of production comes to 45%.
Total profit is $14.35 ($20 - $5.65). Considering the 10% royalty, P/O
split and taxation, the contractor share of *profits* is $3.46. Contractor take
therefore is 24% ($3.46/$14.35). Government take is 76%.

Table 1–12 summarizes the respective "takes" from the cash flow
projection Government take is 76%. The quick-look estimate from the
flow diagram yields the same result—as it should. The only difference

between the flow diagram and the detailed cash flow projection in regard to this estimate of take is that the bonus ($5 MM) was ignored in the flow diagram. In the context of a 100-MMBBL field, the bonus is insignificant. Had there been a significant difference between the take calculations between the cash flow model and the flow diagram, this would indicate a possible problem in either the model or the flow diagram.

Table 1–12 Cash Flow Model Summary and Analysis

When the time value of money (present value discounting) is factored-in, government take always goes up. In this case it goes from 76% undiscounted to 86.5% at 12.5% DCF.

Gross Revenues	$2,000,000	
Total Costs	-565,000 (28.25%)	
Total Profit	$1,435,000	
Bonus	- 5,000	
Royalties	- 200,000	
Government Share Profit Oil	- 741,000	
Income Tax	- 146,700	$1,092,700 (Gvt. Take)
Company Cash Flow	$342,300	
Company Take		24% ($342,300/1,435,000)
Government Take		76% (Undiscounted)
Government DCF (12.5%)	$390,612	
Company DCF (12.5%)	$ 60,736	
Government Take Discounted (12.5%)		**86.5%** $390,612/(390,612+60,736)

Effective Royalty Rate and Access to Gross Revenues

Another index that adds dimension to the *take* statistics' *effective royalty rate* (ERR) is also referred to as *revenue protection*. The complement of ERR is an important oil company viewpoint—access to gross revenues (AGR). This system has a simple 10% royalty, but royalties are not the only mechanisms that create a royalty effect.

The ERR/AGR calculation requires a simple assumption—that expenditures and/or deductions in a given accounting period relative to gross revenues are unlimited. Therefore C/R is at its maximum (i.e., *saturation*), and deductions for tax calculation purposes yield zero taxable income if appropriate for the fiscal system. Situations like this can occur in the early stages of production, with marginal or submarginal fields, or at the end of the life of a field. The object of the exercise is to test the limits of the system. This provides the ERR/AGR indices—the absolute minimum share of revenues a government might expect and the theoretical absolute maximum revenue an oil company might expect.

The ERR is defined as the minimum share of gross revenues a government will get in any given accounting period for a given development, excluding National Oil Company (NOC) share through its working interest participation. AGR, the complement of ERR is the maximum share of revenues the contractor or consortium (including the NOC) can receive in any given accounting period.

The ERR and AGR indices provide important analytical perspective. In an R/T system with no C/R limit, the royalty is the only component of the ERR. It is the only mechanism that provides the government revenue protection. AGR is therefore limited only by the royalty. In most R/T systems in any given accounting period, there is no limit to the amount of deductions a company may take and companies can be in a no-tax-paying position. But this can occur with a PSC as well, as shown in Figure 1–13. Here the limits of the system are tested in the first two years of production—years 3 and 4. C/R is saturated and there is no taxable income. Of the $122 MM revenues generated in this year, the government receives $12.2 MM in royalties and $29.28 MM in P/O (no taxes). This amounts to 34% of total revenues—ERR is equal to 34%. The contractor AGR is equal to 66%. This consists of two components: cost oil and P/O.

Under a PSC with a C/R limit, the NOC is guaranteed a share of P/O because a certain percentage of production is always forced through the P/O split.

Royalties and C/R limits guarantee the government a share of revenues or producton, regardless of whether true economic profits are generated. At the end of the life of an oilfield, the government-guaranteed share of gross revenues takes on the characteristics of a pure royalty.

The take calculations in Table 1–13 illustrate the regressive effect of the 10% royalty as well as the C/R limit. The step-by-step allocation of revenues under high-, low-, and zero-cost cases is shown here with government take decreasing with increased profitability.

Table 1–13 Different Perspectives on the Example PSC

> **This example, similar to that of the R/T system in Table 1–7, compares three scenarios with varying costs and profits. Like the R/T system, this PSC system is regressive. But in this case, it is a function of the royalty and the cost recovery limit.**

Gross Revenues = 100% Full Cycle
Royalty = 10%
Cost Recovery Limit = 50%
Gvt. P/O = 60%
Taxes = 30% CIT

Costs as a percentage of Gross Revenues—Three Scenarios:

60% - High-cost case
↓
30% - Low-cost case
↓
0% - at the margin
↓

High-Cost Case	Low-Cost Case	Zero-Cost Margin	
A	**B**	**C**	← [see Fig. 1–14]
100.00%	100.00%	100.00%	**Gross Revenues**
-10.00	- 10.00	- 10.00	**Royalty**
90.00	90.00	90.00	Net Revenues
- 50.00*	-30.00	- 0.00	**Total Cost Recovery**
40.00	60.00	90.00	Profit Oil
-24.00	-36.00	-54.00	**Gvt. Share 60%**
16.00	24.00	36.00	Contractor Profit Oil
-10.00			Unrecovered Costs
6.00	24.00	36.00	Taxable income
-1.80	-7.20	-10.80	**Income Tax 30%**
4.20	16.80	25.20	Contractor cash flow
10.5%	**24.0%**	**25.2%**	**Contractor Take**
89.5%	**76.0%**	**74.8%**	**Government Take**

* Costs = 60%; but cost recovery limit = 50%.

In the high-cost case, costs exceed the C/R limit, yet there are sufficient revenues to allow full-C/R. However only part of the C/R comes through the C/R mechanism itself. Additional recovery comes through the company share of P/O.

A graphical representation is shown in Figure 1–14. It shows the modest regressive effect of the royalty and the combined effect of royalty and C/R limit when costs relative to gross revenues exceed the limit. The basic equations for contractual systems such as PSCs are found in Table 1–14.

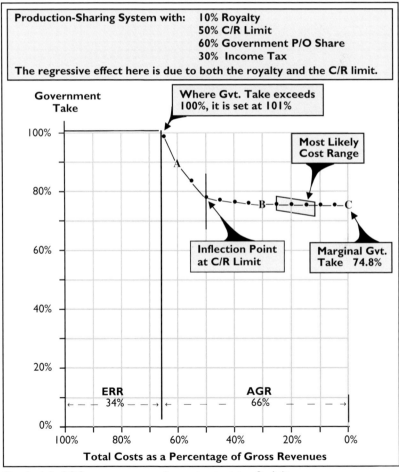

Fig. 1–14 PSC government take vs. project profitability

Table 1–14 Basic Equations—Contractual Systems

Gross Revenues	= Total oil and gas revenues

Cost Recovery
Cost Oil
= Operating costs
+ Intangible capital costs *
+ DD&A
+ Investment credits (if allowed) **
+ Interest on financing (if allowed)
+ Unrecovered costs carried forward
+ Abandonment cost provision

Profit Oil
= Gross production − Royalty oil − Cost oil

Contractor Profit Oil
= Contractor profit oil
+ Contractor cost oil

Contractor Entitlement
= Profit oil
* Contractor percentage share

Government Profit Oil
= Profit oil
* Government percentage share

Taxable Income
= Gross revenues
− Royalties
− Intangible capital costs *
− Operating costs
+ Investment credits (various incentives)
− Government profit oil
− Abandonment cost
− DD&A
− Bonuses ***
− TLCF

Net Cash Flow
(after-tax)
= Gross revenues
− Royalties
− Tangible capital costs
− Intangible capital costs *
− Operating costs
+ Investment credits
− Bonuses
− Government profit oil
− Taxes

* In many systems, no distinction is made between *operating costs* and *intangible capital costs* and both are *expensed*.
** As a general rule, investment credits are cost recoverable but they are not tax deductible.
*** Another general rule: bonuses are often not cost recoverable but they are tax deductible.

Risk Service Contracts

The service contract concept is primarily based upon a simple formula: the contractor is paid a cash fee for performing the service of producing mineral resources. All production belongs to the government.

The contractor is usually responsible for providing all capital associated with exploration and development of petroleum resources. In return, if exploration efforts are successful, the government allows the contractor to recover those costs through sale of the oil or gas and pays the contractor a fee that is usually based on a percentage of the remaining revenues. This fee is often then subject to taxes, and it is very similar to a PSC.

There is quite a variety of service contract systems. The terminology is widely accepted but rather inappropriate. The oil service industry would hardly recognize the service contracts found in the upstream end of the business. To refer to an *exploration agreement* (where the oil company puts up all the capital and risks loosing it all) as a *service agreement* is an obvious misnomer. But this is what it is called. The added term *risk* is clearly an improvement.

Because the contractor does not get a share of production, such terms as *production sharing* and *profit oil* are not appropriate even though the arithmetic will often carve out a share of revenues in the same fashion that a PSC shares production.

Usually when the term *service contract* is used, it is understood that it is a *risk-service contract*. However, sometimes the term *risk contract* is also used. Under risk-service agreements, the contractor bears all the risk but has the potential of profits.

Pure Service Contract

Pure (non-risk) service contracts are where the contractor carries out exploration and/or development work on behalf of the host country for a fee. All ordinary exploration risk is borne by the state. This kind of arrangement is more characteristic of the Middle Eastern region, where the state often has substantial capital but seeks only expertise.

Pure service agreements, rare as they are, can be quite similar to those arrangements found in the oil service industry. The contractor is paid a fee for performing a service. In the late 1950s, the Argentine government under President Arturo Frondizi negotiated a number of service contracts known as *The Frondizi Contracts*. These contracts were negotiated with oil companies for drilling services, development services, and medium-risk exploration services. These companies included Kerr McGee, Marathon, Shell, Esso, Tennessee Gas Transmission, Cities Service, Amoco, and Union Oil. The drilling service contracts were pure service arrangements whereby the contractor was paid on a footage basis (while drilling) and an hourly basis (for testing and completion operations). The payment was usually a combination of dollars and pesos.

Many service agreements are identical to PSCs in all but the method of payment—either production sharing or profit sharing. However, many service agreements have unique contract elements that are used in calculating the service fee.

R Factor Based Systems

R factors (sometimes called *Factor R*) are based on an ancient and fundamental formula—payout.

Typically the contract will stipulate that for example, a tax rate may be subject to a factor R, and several *thresholds* will be established either through negotiation or they may be statutory. And R, which stands for *ratio*, will be a function of X divided by Y (X/Y). X is defined as the accumulated receipts actually received by the contractor less tax. Y is defined as the accumulated capital expenditures (Capex) and operating expenditures (Opex). The factor R is calculated in each accounting period; and once a threshold is crossed, then the new tax rate will apply in the next accounting period.

$$R = \frac{X}{Y}$$

Where: X = Contractor cumulative receipts (after-tax) [Cumulative net revenue actually received by the contractor equals *turnover* (gross revenues) for all tax years less taxes paid.]

Y = Contractor cumulative expenditures [Total cumulative expenditure, exploration, and appraisal expenses, development expenditures and operating costs actually incurred by contractor from date contract is signed.]

At payout, X is equal to Y. This yields an *R* factor of 1.

The most common use of the *R* factor is found in the Tunisian and Peruvian contracts. In these contracts, the definitions are virtually identical; *R* factor = Accrued Net Earnings/Accrued Total Expenditures. In Tunisia oil and gas royalties, taxes and government participation are all based upon the *R* factor. An example of the *R* factor-based royalty is shown as follows:

Royalty Rate

R Factor	Oil	Gas
< .5	2%	2%
.5 - .8	5	4
.8 - 1.1	7	6
1.1 - 1.5	10	8
1.5 - 2.0	12	9
2.0 - 2.5	14	10
2.5 - 3.0	15	11
3.0 - 3.5	15	13
3.5 +	15	15

In this example, once the contractor has reached halfway to payout ($R = 0.5$), the royalty rate starts to increase. In some respects, it is similar to a ROR contract. The ROR contracts trigger on internal rate of return (IRR) thresholds; yet as part of the language of the industry, they are referred to as *ROR* systems. Yet these are called *ROR* systems not *IRR* systems. These are discussed next.

Figure 1–15 illustrates the effect an *R* factor can have on project economics. The results on contractor project IRR are shown as costs and oil prices vary. Costs and prices are the most sensitive factors in project economics. They have the largest impact and with an *R* factor both are accounted for simultaneously. The *R* factor deals with all variables that impact project economics. The sensitivity is shown by the slope of the lines. The *R* factor has a dampening effect. Contractor potential upside from price increases is diminished, but the downside is also protected. It is the same with costs. If costs are relatively higher, the *R* factor mitigates the negative impact to the contractor. If costs are lower, both the contractor and the government benefit.

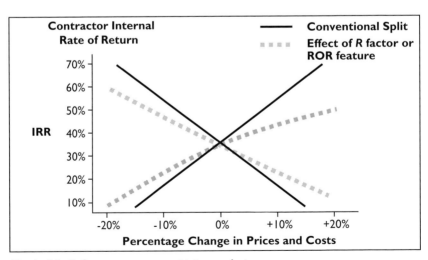

Fig. 1–15 *R factor system sensitivity analysis*

ROR Systems

Some countries have developed progressive taxes or sharing arrangements that are based upon project ROR. The effective government take increases as the project ROR increases. In order to be truly progressive, the sliding scale taxes and other attempts at flexibility must be based upon profitability. Most contracts have progressive elements, but they are usually based upon levels of production instead of a direct measure of profitability. Production levels are often a good proxy for profitability but that is all. There are many other factors that influence project profitability and that is why ROR contracts are structured the way they are.

ROR contracts directly take into account such things as:

- Production profiles
- Oil & gas prices
- Costs
- Cost of capital
- Timing

The ROR approach is characterized by a modest royalty and tax that the state receives; but the state receives no other funds until the oil company has recovered the initial financial investment plus a predetermined threshold ROR. Theoretically this ROR would represent a minimum rate to encourage investment. The government share is calculated by accumulating the negative net cash flows and compounding them at the threshold rate until the cumulative value becomes positive.

Example ROR System

The example system summarized in Table 1–15 is typical of the classic ROR formula. Under this system, the government receives a 10% royalty. A basic income tax of 40% is levied if the contractor has generated taxable income.

Table 1–15 Example ROR System

Summary of Commercial Terms	
Bonus	**$5 MM**
Tax deductible but not cost recoverable.	
Royalty Rate	10%
Basic Income Tax	40%
Resource Rent Tax	50%
Triggered @ 30% IRR	
Summary of Economic Analysis	
Government Take	60.7%
(Undiscounted)	
Government Take	64.2%
@ 12.5% DCF	
Effective Royalty Rate (ERR)	10%
Access to Gross Revenues (AGR)	90%
Entitlement Index	56%
Savings Index	28¢

There is an additional tax levied, the resource rent tax, if the contractor ROR exceeds 30%. This is determined by compounding and accumulating the negative net cash flows at a rate of 30%. This is called *compound uplifting*. Once the cumulative net uplifted cash flow becomes positive, the additional 50% resource rent tax kicks in. This is the hallmark of an ROR system. It is also called a *trigger tax*. Reaching a minimum ROR (in this case 30%) triggers the tax. Government take in this system is summarized in Table 1–16. The basic structure of this example ROR contract is illustrated in Figure 1–16.

Table 1–16 ROR System Cash Flow Model Summary and Analysis

Gross Revenues	$2,000,000	
Total Costs	-565,000	(28.25%)
Total Profit	$1,435,000	
Bonus	- 5,000	
Royalties (10%)	- 200,000	
Basic Income Tax (40%)	-492,000	
Resource Rent Tax (50%)	- 174,758	$871,758 (Gvt. Take)
Company Cash Flow	$563,242	
Company Take	39%	($563,242/1,435,000)
Government Take	61%	(Undiscounted)
Government DCF (12.5%)	$289,642	
Company DCF (12.5%)	$ 161,706	
Government Take		
Discounted (12.5%)	64%	
(289,642/(161,706+289,642)		

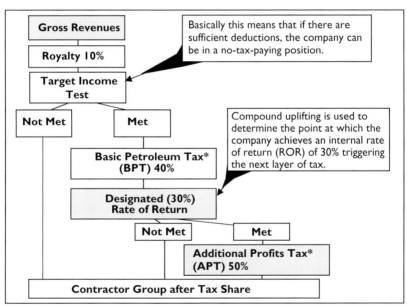

Fig. 1–16 Example system: R/T based ROR system

Calculation of cash flow the example ROR system is shown in Table 1–17 and Table 1–18. It outlines the basic ROR system elements with a detailed explanation of the calculations involved in arriving at year-by-year cash flow. In this example, the royalty is 10%, and the basic income tax is 40%. A 30% uplift is applied on the accumulated negative net cash flows. Once the cumulative balance of net cash flow becomes positive (the point at which the IRR of 30% has been reached), an additional 50% resource rent tax is imposed. Table 1–19 shows that in Year 10 of the cash flow model when $38,452 M has been generated, the IRR is 30%.

Table 1–17 Example ROR System Cash Flow Projection Field X Development Feasibility Study

Year	Annual Oil Production (MBBLS)	Oil Price ($/BBL)	Gross Revenues ($M)	Royalty 10% ($M)	Net Revenue ($M)	Capital Costs ($M)	Op. Costs ($M)	Depreciation ($M)	Bonus ($M)	TLCF ($M)
	A	B	C	D	E	F	G	H	I	J
1	0	$20				30,000			5,000	0
2	0	$20				40,000				5,000
3	578	$20	11,560	1,156	10,404	100,000	3,156	34,000		5,000
4	6,100	$20	122,000	12,200	109,800	60,000	16,200	46,000		31,752
5	9,420	$20	188,400	18,840	169,560	70,000	22,840	60,000		
6	12,400	$20	248,000	24,800	223,200		28,800	60,000		
7	10,850	$20	217,000	21,700	195,300		25,700	60,000		
8	9,494	$20	189,880	18,988	170,892		22,988	26,000		
9	8,307	$20	166,140	16,614	149,526		20,614	14,000		
10	7,269	$20	145,380	14,538	130,842		18,538			
11	6,360	$20	127,200	12,720	114,480		16,720			
12	5,565	$20	111,300	11,130	100,170		15,130			
13	4,869	$20	97,380	9,738	87,642		13,738			
14	4,261	$20	85,220	8,522	76,698		12,522			
15	3,728	$20	74,560	7,456	67,104		11,456			
16	3,262	$20	65,240	6,524	58,716		10,524			
17	2,854	$20	57,080	5,708	51,372		9,708			
18	2,498	$20	49,960	4,996	44,964		8,996			
19	2,185	$20	43,700	4,370	39,330		7,370			
20										
Total	100,000		2,000,000	200,000	1,800,000	300,000	265,000	300,000	5,000	565,000

Year	Taxable Income ($M)	Basic Income Tax 40% ($M)	Net Cash Receipts ($M)	Amount Brought Forward	Amount Carried Forward	Rent Tax Base ($M)	Resource Rent Tax 50%	Contractor Cash Flow	
								Undiscounted	12.5% DCF
	K	L	M	N	O	P	Q	R	S
1	(5,000)	0	(35,000)		(35,000)			(35,000)	(32,998)
2	(5,000)	0	(40,000)	(45,500)	(85,500)			(40,000)	(33,522)
3	(31,752)	0	(92,752)	(111,150)	(203,902)			(92,752)	(69,094)
4	15,848	6,339	27,261	(265,073)	(237,812)			27,261	18,051
5	86,720	34,688	42,032	(309,155)	(267,123)			42,032	24,740
6	134,400	53,760	140,640	(347,260)	(206,620)			140,640	73,582
7	109,600	43,840	125,760	(268,606)	(142,846)			125,760	58,486
8	121,904	48,762	99,142	(185,700)	(86,558)			99,142	40,984
9	114,912	45,965	82,947	(112,525)	(29,578)			82,947	30,479
10	112,304	44,922	67,382	(38,452)		28,931	14,465	52,917	17,284
11	97,760	39,104	58,656			58,656	29,328	29,328	8,515
12	85,040	34,016	51,024			51,024	25,512	25,512	6,584
13	73,904	29,562	44,342			44,342	22,171	22,171	5,086
14	64,176	25,670	38,506			38,506	19,253	19,253	3,926
15	55,648	22,259	33,389			33,389	16,694	16,694	3,026
16	48,192	19,277	28,915			28,915	14,458	14,458	2,329
17	41,664	16,666	24,998			24,998	12,499	12,499	1,790
18	35,968	14,387	21,581			21,581	10,790	10,790	1,374
19	31,960	12,784	19,176			19,176	9,588	9,588	1,085
20									
Total		492,000	738,000			349,516	174,758	563,241	161,706

Table 1–18 Example ROR System Cash Flow Projection Government Cash Flow

Year	Bonuses ($M)	Royalty 10% ($M)	Basic Income Tax 40% ($M)	Resource Rent Tax 50% ($M)	Government Cash Flow ($M)	
					Undiscounted	12.5% DCF
	I	D	L	Q	T	U
1	5,000		0		5,000	4,714
2			0		0	0
3		1,156	0		1,156	861
4		12,200	6,339		18,539	12,276
5		18,840	34,688		53,528	31,506
6		24,800	53,760		78,560	41,102
7		21,700	43,840		65,540	30,480
8		18,988	48,762		67,750	28,007
9		16,614	45,965		62,579	22,995
10		14,538	44,922	14,465	73,925	24,146
11		12,720	39,104	29,328	81,152	23,561
12		11,130	34,016	25,512	70,658	18,235
13		9,738	29,562	22,171	61,471	14,101
14		8,522	25,670	19,253	53,445	10,898
15		7,456	22,259	16,694	46,409	8,412
16		6,524	19,277	14,458	40,259	6,486
17		5,708	16,666	12,499	34,873	4,994
18		4,996	14,387	10,790	30,173	3,841
19		4,370	12,784	9,588	26,742	3,026
20						
Total	5,000	200,000	492,000	174,758	871,758	289,642

A)	**Production Profile** Thousands (M) barrels/year	K) **Taxable Income** $= (C - D - G - H - I - J)$
B)	**Crude Price**	L) **Income Tax 40%** $= (K * .40)$
C)	**Gross Revenues** Thousands of dollars ($M)	M) **Net Cash Receipts** $= (K - L)$
D)	**Royalty 10%** $= (C * .10)$	N) **Amount Brought Forward** $=$ (O uplifted)
E)	**Net Revenues** $= (C - D)$	O) **Amount Carried Forward** $= (M + N)$
F)	**Capital Costs**	P) **Res. Rent Tax Base** $=$ (O if > 0)
G)	**Operating Costs** (Expensed)	Q) **Resource Rent Tax 50%** $= [P * .50]$
H)	**Depreciation** of Capital Costs (5-year SLD)	R) **Company Cash Flow** $= (M - Q)$
I)	**Signature Bonus**	T) **Government Cash Flow** $= (I + D + L + Q)$
J)	**TLCF** $=$ (See Column K)	

Table 1–19 IRR of Net Cash Receipts at Trigger Point

Year	Net Cash Receipts ($M)	Amount Brought Forward	Net Cash Receipts at Trigger Point	
			Undiscounted	30% DCF
	M	N		
1	(35,000)		(35,000)	(30,697)
2	(40,000)	(45,500)	(40,000)	(26,986)
3	(92,752)	(111,150)	(92,752)	(48,135)
4	27,261	(265,073)	27,261	10,883
5	42,032	(309,155)	42,032	12,907
6	140,640	(347,260)	140,640	33,222
7	125,760	(268,606)	125,760	22,851
8	99,142	(185,700)	99,142	13,857
9	82,947	(112,525)	82,947	8,918
10	67,382	(38,452)	38,452	3,180
11	0			
12				
Total			463,482	0

In Year 10, only the Net Cash Receipts required to balance out the Amount Brought Forward of $38,452 are used for this analysis. The discount rate that yields a present value of $0 (zero) is 30%—the same as the uplift rate. Thus at the point that the Amount Brought Forward goes from negative to positive, the company IRR is equal to the uplift rate.

Critics of the ROR concept complain that these contracts are too restrictive, that the uplift (ROR) places an unreasonably low ceiling on upside potential. This is not a fair criticism. Most of the criticism about ROR systems, including the claim that they inspire gold plating, are based on false logic. Generally speaking, companies would likely develop an oilfield in the same way under a ROR-based system as any other system.

The ROR concept was first employed in Papua New Guinea (PNG). Other countries that use this kind of tax are Australia (the only other system in Southeast Asia), Liberia, Equatorial Guinea, and Tanzania (and numerous other systems in Africa), and Kazakhstan and Azerbaijan (and many other republics of the Former Soviet Union or FSU).

Most other ROR systems have multiple thresholds instead of a single threshold like the one in the example that follows.

Notice in the ROR system cash flow model that the R factor at the end of the life of the field is only 2 (actually 1.996). Why was it not more?

Oil prices are not bad at $20 per barrel (held constant).

Costs are fairly low:
Capex is only $3/BBL
Opex is only $2.65/BBL

Total costs as a percentage of gross revenues is only 28.3%.

Profitability is fairly good, and of those profits, the government only takes 60.7%, which is not bad—government take is 60.7%. Yet the R factor is only 2. Economics must be fairly robust for a company to get all of its money (both Capex and Opex) back plus 100%. This provides an important guideline in evaluating or designing systems with R factors. Figure 1–17 depicts the correlation between R factors and internal rate of return (the basis of ROR systems).

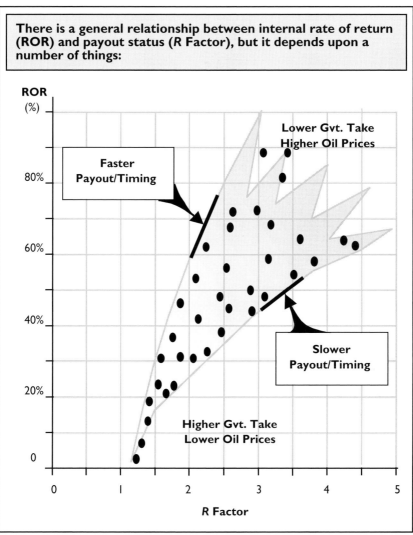

There is a general relationship between internal rate of return (ROR) and payout status (R Factor), but it depends upon a number of things:

Fig. 1–17 R factors vs. ROR systems

JOINT VENTURES

Joint ventures are a common mode of doing business in the international oil industry. Many companies *partner up* for large-scale or for high risk ventures in order to diversify—good risk management. These joint operations between industry partners differ from the government-contractor relationships, which are also joint ventures but are normally referred to as government participation.

Some contracts provide an option for the NOC to participate in development projects. These government participation clauses usually stipulate that the NOC has the right to join in development as a working interest partner sometimes after paying a prorated share of exploration costs.

The key aspects are:

- What percentage participation?

- When does the government back in?

- How much participation in management?

- What costs will the government bear?

- How does government fund its portion of costs?

Some of the proposed Russian Joint Ventures (JV) are characterized by a 100% carry for the production association partner through development including operating costs as well. This is an extreme example of government participation. However, most of the Russian JVs deal with proved, well-delineated reservoirs. The exploration risk aspect is greatly diminished.

Under most government participation arrangements, the contractor bears the costs and risks of exploration; and if there is a discovery, the government backs-in for a percentage. In other words, the government is carried through exploration. This is fairly common and is automatically assumed whenever some percentage of government participation is quoted. However, as with everything else, there are a lot of variations on this theme. In Colombia, the government is carried through exploration and two delineation wells but will reimburse the company for successful exploration wells. Some (very few) countries will reimburse all exploration costs if the efforts are successful. Usually, exploration costs are recoverable or deductible, so the government pays for a portion of them indirectly anyway.

In cases where the government is carried through exploration, and backs in upon discovery, the impact of government participation behaves like a capital gains tax. In the extreme cases like Russia where the contractor pays all rehabilitation, development, and operating costs, the government share in joint-venture profits behaves like an added layer of taxation. Where the government actually pays its share of costs, the government share of profits is not the same as a tax on income. In either case, almost all forms of government participation reduce the potential reward that may result from exploration. This must be factored into the risk equation.

GLOBAL MARKET

Governments are increasingly aware of their position in the global market for exploration acreage or rehabilitation projects. The international market for drilling funds and technology is increasingly competitive and sophisticated. Governments are becoming better aware of what the market can bear and know how to adjust their terms to compete with countries that have greater geological potential. This is illustrated in Figure 1–18.

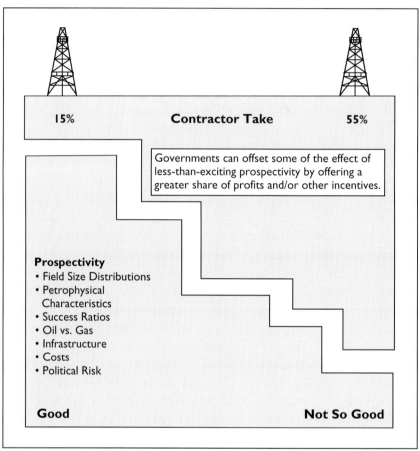

Fig. 1–18 Creating a level playing field

There are still sedimentary basins with promising geological conditions that are virtually unexplored. These basins are usually more geologically complex, or are located in relatively deeper waters. Yet most of the basins with drilling and production history have a long way to go in terms of exploration and development maturity by comparison to the United States.

Most regions are at a stage that existed from 20 to 40 years ago in the United States. The total number of wells drilled in the United States now exceeds 3 million. This is two orders-of-magnitude greater than the 30,000 odd wells that have been drilled in the Asia Pacific region (excluding China) and nearly 4 times as many as have been drilled in Russia. If the industry drilled 50,000 wells per year in Russia, it would take 40 years to catch up to where the United States is now.

Much of the U.S. oil patch is super-mature. Of the nearly 600,000 producing oil wells in the United States, more than two thirds are stripper wells—producing less than 10 BOPD. The average production rate for these 400,000 odd stripper wells is barely more than 2 BOPD. This is part of the problem in the United States. There is still geological potential in areas of the United States, but most of these areas are off-limits to the oil industry. When explorers go overseas, they focus on geopotential and how that potential balances with fiscal terms and the cost of doing business.

When evaluating fiscal terms, the focus is on how profits are divided, meaning what is the contractor take. Figure 1–18 illustrates the effective trade-off in terms of the contractor take and geopotential, costs and the other key factors that influence business decisions. The split in most countries ranges from just under 15% to over 55%. Beyond these extremes are the exceptions that are becoming more and more rare.

The calculation of contractor take provides the answer to what should be the first question asked about a fiscal system. The second question should address the limits on C/R such as the ERR. The rate at which costs are recovered can have a huge impact on economics, particularly marginal field developments. On the larger more profitable fields, economic results are not so sensitive to C/R limits or ERR.

There are other, less direct limits or influences on C/R than the typical PSC C/R limit. They include:

- Ringfencing (most countries have ringfencing)
- Depreciation rates
- Royalties can limit revenues available for C/R or deductions
- Contractor share of P/O or profits enhances the C/R process

COMMENTARY

The exploration business starts with geology. Fiscal terms are of secondary but critical importance. Generalizations are often made about the superiority of concessionary systems over PSCs from the oil company point of view. A better, more appropriate generalization is one that compares fiscal systems on the basis of contractor take.

Contractor take (or government take) provides the best single means of characterizing a given fiscal system. It is not a perfect statistic and weaknesses are discussed in later chapters. The most important aspect after contractor take is the ERR companion statistic. While most PSCs have a formal C/R limit, concessionary systems have limits too. A royalty imposes an indirect limit, and long-term depreciation schedules do too.

The issue of ownership is most important to companies who have refineries that need to be supplied. Beyond that, from a financial point of view, the differences between systems begin to disappear.

Governments are becoming increasingly aware of their position in the global marketplace, and competition is warming up.

Changes in fiscal terms cannot cure all problems, but some countries have a lot of room for enhancing their terms. Changes are taking place at a furious pace.

2

State-of-the-Art in Petroleum Fiscal System Analysis

One of the most pressing concerns of most governments dealing with petroleum or mining licenses is the division of profits—government take.

In that regard, the guiding lights in petroleum fiscal system design are efficiency, flexibility, and, of course, global competition. The objective is to create a system that efficiently captures economic rent (profits) yet is flexible enough to encompass a variety of conditions and possible outcomes. The process by which governments attempt to determine "what the market can bear" is sometimes referred to as fiscal marksmanship. This can be a heavy burden for a government agency. Some countries avoid the risk of getting it wrong by allowing substantial elements to be open for competitive bidding or negotiation. The ultimate test is the number of licenses awarded.

There are seven fundamental criteria that capture the essence of the commercial aspects of a fiscal system. These are:

- The division of profits—contractor/company take
- Royalties
- Cost recovery limits
- AGR
- Government carry
- Ringfencing
- Crypto taxes, including:
  - —Value-added taxes
  - —Training obligations
  - —Rental payments for acreage
  - —Mandatory currency conversions
  - —National employment quotas
  - —Bonuses (sometimes included in take statistic)
  - —Import duties
  - —Any items that are difficult to collapse into the take statistic

THE DIVISION OF PROFIT—
CONTRACTOR TAKE

The most critical factor in government competitive strategy is the division of profits. The division of profits is also a key element in the analysis of any fiscal system or drilling venture. Unfortunately, the terminology surrounding this concept is diverse. Half a dozen different terms are used to describe the division of profits. Furthermore, there are conceptual differences regarding the characterization of the division of profits.

Much of the existing terminology began with the Indonesian fiscal system. In early vintage Indonesian PSCs with no royalty and no taxes, the respective shares of profit oil (equity oil) were the same as the division of profits. Thus the famous 85/15% *equity split* in Indonesia. After taxes were introduced in 1978, the terminology relating to the division of profits technically became the *after-tax equity split*, which was still 85/15%.

These terms are still used in Indonesia, but they are not quite appropriate for most other fiscal systems. First of all, R/T systems (concessionary systems) have no profit oil split or equity oil. For many production-sharing systems, the profit oil split does not represent the *division of profits* as it did in Indonesia because many PSCs have a royalty.

Regardless of the type of fiscal system or the terminology, the bottom line is how to divide profits and, to a lesser extent, how to recover costs. The industry is particularly sensitive to certain forms of government take such as bonuses and royalties that are not based on profits. Royalties are of particular concern to the industry because the rate base for royalties is gross revenues. Government participation is usually in the form of a risk-free carry through exploration. It adds an interesting dynamic to fiscal system analysis.

The nonprofit related elements of government take such as royalties and bonuses are regressive—the lower the project profitability, the higher the effective tax rate. The further downstream from gross revenues a government levies taxes, the less regressive the system becomes. This is becoming more common. Royalties are being reduced in favor of profits-based taxes. This has advantages for both governments and the petroleum industry. However, there will always be governments that prefer some royalty. Royalties provide a guarantee that the government will benefit in the early stages of production and, of course, cost recovery limits will do this too.

The division of profits is most commonly referred to these days as *contractor* and *government take*, which are expressed as percentages. Contractor take is the percentage of profits to which the contractor or oil company (the terms here are used interchangeably) is entitled. Government take is the complement of that. The list that follows clarifies some of the terminology.

DIVERSITY OF TERMINOLOGY— THE DIVISION OF PROFITS

"The Split"

Company/Contractor	Government
Contractor equity split	Government equity split
Contractor after-tax equity split	Government after-tax equity split
Contractor take	State take
Contractor take	Government take
Discounted contractor take	Discounted government take
Contractor marginal take	Government marginal take
Net contractor take on the marginal barrel	Net government take on the marginal barrel
Contractor take	Tax take
Contractor take	Government take (state)
Contractor cash margin	Government cash margin

Most of the time these statistics refer to undiscounted profits. It is usually sufficient to characterize the division of profits from the undiscounted perspective rather than to view the division of profits from a present value point of view. In nearly every fiscal system, the government share of discounted profits is greater than the government share of undiscounted profits. For example, the famous Indonesian 85/15% split with the DMO and investment credit (IC) factored-in is closer to an 87/13% split (undiscounted). With a 15% discount rate, this split approaches 93/7% for full-cycle exploration economics. The formulas for calculating take are summarized in Table 2–1.

Table 2–1 Formulas for Calculating Take

Net cash flow ($) (Economic profits) (Operating income)	=	Cumulative gross revenues less cumulative costs over the life of a field or license
Government net cash flow ($)	=	All government receipts from royalties, taxes, bonuses, production, or profit sharing (excludes government working interest share of net cash flow)
Government take (%)	=	Government net cash flow divided by total net cash flow
Contractor take ($)	=	Contractor net cash flow
Contractor take (%)	=	Contractor net cash flow divided by total net cash flow
Contractor take (%)	=	1 - government take (%)

Note: State take is the same as Government take except that it includes the government share of net cash flow that would result from a working interest ownership (as part of the contractor group).

Contractor take provides an important point of comparison between one fiscal system and another, as well as between contracts. This statistic focuses exclusively on the division of profits and correlates with reserve values, field size thresholds, and other measures of *relative* economics. How tough are the fiscal terms? The answer to this question is found in the contractor take statistic. Governments that have excellent potential, infrastructure, and political conditions can extract a greater percentage of profits, yet countries with less potential can compete as well. The division of profits/contractor take for profitable field developments ranges from 15% to 55%. This captures approximately 90% of the systems in the world. Figure 2–1 shows a representative sampling of the universe of fiscal systems in the world. In Figure 2–1, the government carry is ignored.

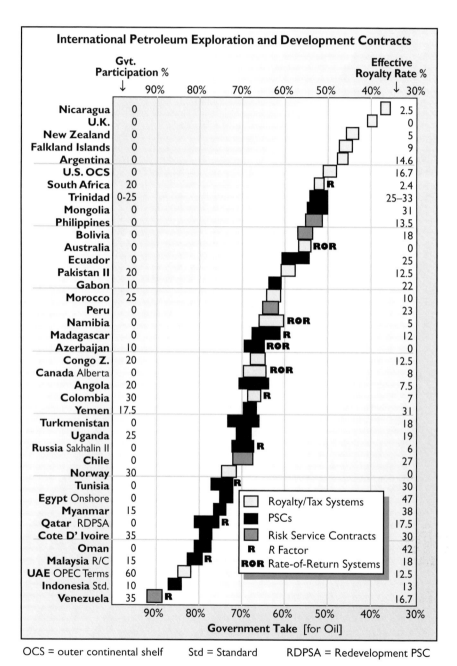

Fig. 2–1 Comparison of fiscal terms

The best way to calculate take requires detailed cash flow analysis. Once a cash flow projection has been performed, the division of profits over the life of the project can be evaluated. However, there is a common method for estimating government/contractor take without detailed cash flow modeling. There are limitations to this quick-look approach, but 95% of the time, estimates of contractor/government take provide valuable information.

The examples in Table 2–2 show different calculations of take for a system with a simple 10% royalty, a 40% income tax, and a 50% government carry. Capital and operating costs over the life of the field range from 30 to 50% of gross revenues. These are undiscounted takes. Using a discount rate of 15%, the split in Case II is 70/30% in favor of the government in an exploration economics scenario.

Table 2–2 Same Fiscal System—Different Takes

Case I	Case II	Case III	Case IV	
"Zero" Cost	Low Cost	High Cost	Low Cost	
100.00	100.00	100.00	100.00	Gross revenues
-10.00	-10.00	-10.00	-10.00	**Royalty 10%**
	90.00	90.00	90.00	Net revenues
	-30.00	-50.00	-30.00	Costs
90.00	60.00	40.00	60.00	Taxable income
-36.00	-24.00	-16.00	-24.00	**Income tax 40%**
54.00	36.00	24.00	36.00	Contractor group net cash flow
			-18.00	**Government carry 50%**
			18.00	Contractor net cash flow
54.0%				Contractor marginal take
46.0%				Government marginal take
	51.4%*	48.0%	25.7%	**Contractor take**
	48.6%	52.0%		Government take
			74.3%	State take

* For example: The contractor share of profits amounted to 36% of gross revenues. Total profits amounted to 70% of gross revenues (100% – 30% costs). Therefore, the contractor take (or share of profits) amounts to 36/70 or 51.4%.

Sometimes when a "Government take" statistic is quoted, it represents a percentage of revenues (not profits. In Case IV for example, the royalty, income tax, and Government carry (participation) yields a government share of revenues of 52% (10% + 24% + 18% respectively).

Contractor *take* under the high cost Case III is 48%, yet under a more profitable environment such as Case II (same fiscal system) the contractor's *share* of profits increases to 51.4%. Conversely, the lower the profitability, the higher the effective tax rate. This is a regressive fiscal structure. The royalty is based on gross revenues. Most governments would prefer to have a higher take under conditions of greater profitability. However, this will not happen if the system has a royalty. Most governments would rather live with a measure of regressiveness than not have the protection of a royalty. Royalties guarantee that the government will get a share of gross revenues regardless of the profitability of a venture.

Marginal Take

The *net take on the marginal barrel* or *marginal take* approaches are a slightly abstract view of the *take* concept that effectively ignores the cost element. The typical approach is illustrated in Case I in Table 2–2. This approach distorts somewhat the actual division of profits and neutralizes the regressive effect of royalties and royalty equivalent elements. Notice for example, that in Case I the royalty represents 14.28% of the total profits. By contrast, the same 10% royalty represents 20% of the total profits in Case III.

Marginal take will almost always be equal to or greater than ordinary estimates or calculations of take. For those fiscal systems with no royalty, marginal take equals take. For example in the United Kingdom where this terminology is used most often, there is no royalty. Contractor take is 67%. Marginal take is also 67%.

Government vs. State Take

The government take calculation as illustrated in Cases I, II, and III effectively ignores government participation. (The government risk-free carry through exploration is also known as *government back-in* where/when the government-owned NOC has the option of taking up a working interest share of the project if a discovery is made). State take, on the other hand, as some analysts use the term, directly includes government participation (the government share of profits resulting from the working interest obtained when the government *backs-in*, which is the hallmark of the government participation element).

About 40% of the countries in the world have the option of participating as a working interest partner at the point of discovery. This is ordinarily referred to as *government participation* or the *government carry*. For those countries that have this option, the average carried interest is around 30%. The range is generally from 15 to 50%. At the point of discovery (usually plus a delineation well or two—the point of commerciality) the government takes up its working interest share and "pays their way" from that point forward. And, typically, the contractor is reimbursed for the government share of past costs. It adds an interesting dynamic to fiscal system analysis and creates a bit of controversy. There are a number of analysts who believe that the government *carry* should be treated effectively as though it were a tax as in Case IV.

When discussing the division of profits for a given fiscal system or contract, it is important to understand the diversity of terminology as well as the different viewpoints. Unfortunately, there is not a different corresponding or complementary term for state vs. government take—the complement of each is contractor (or company) take. So if someone refers to a contract with a contractor take of X%, the statistic does not by itself indicate how the government carry element was factored in. However, for development threshold field size analysis, development economics, proved developed producing (PDP) reserve value analysis, or production acquisitions, the government participation element has virtually no meaning. Once the government has *backed-in*, the government is just like any other working interest partner from a financial point of view.

People use the term *split* these days almost always as a synonym for *take*, but the terms are not really interchangeable. In current business usage, *take* usually refers to undiscounted profits—the division of profits can look dramatically different from a discounted point of view. If a system has a royalty or royalty equivalent, there will be different *takes* for different degrees of profitability, and the system will be regressive. Royalties are embodied in the contractor/government take statistic, but the royalty rate gives a perspective on the *efficiency* of the system.

ROYALTIES

Determination of the ERR for a given fiscal system is tantamount to determination of the system's *efficiency*. Royalties are by definition regressive, which is the antithesis of efficiency. One aspect of royalties that contributes to their lack of popularity with the industry is that they can prematurely cause production to become uneconomic—to the disadvantage of both industry and government. Royalties are effective though, and they ensure that in any given accounting period, the government will not go away empty-handed. Cost recovery limits perform this same function. (*See* the following section, *AGR*, for a discussion of the combination of royalties and cost recovery limits.)

Royalties typically range from zero to 20%. Anything above 15% is getting excessive. High royalties are inefficient and counterproductive. A 20% royalty on $18 oil is $3.60 per barrel. This makes a huge difference with small field developments and marginal production. A marginal field, for example, may require more than 50% of gross revenues over the life of the field for recovery of capital and operating costs. A 20% royalty in that situation would represent more than 40% of profits. One remedy that is popular is to scale royalties and other fiscal elements to accommodate marginal situations.

COST RECOVERY LIMITS

Cost recovery is the means by which the contractor recoups costs of exploration, development, and operations out of gross revenues. Cost recovery is an ancient concept. Whoever puts up the capital should *at least* get the investment back. The cost recovery mechanism is one of the most common features of a PSC and differs only slightly from the cost recovery techniques used in most R/T (concessionary) systems. In most respects, cost recovery is identical to deductions in calculating taxable income under a concessionary system. Most PSCs place a cap on the amount of revenues the contractor may claim for cost recovery in any given accounting period. Unrecovered costs are normally carried forward and recovered in successive accounting periods. Cost recovery limits, or cost recovery *ceilings* as they are also known, range (if they exist) typically from 40% to 70%. R/T systems normally do not limit the amount of deductions that can be taken.

The cost recovery limits under most PSCs create an added dimension to fiscal system analysis. When the Indonesian government in 1988/89 created the first tranche petroleum (FTP) of 20%, some analysts claimed that it represented the equivalent of a royalty. Others argued that it was nothing more than an 80% cost recovery limit. The fact that there was a debate at all is significant. Cost recovery limits, especially at the end of the life of a field, take on many of the characteristics of a royalty. The best way to view this aspect of a fiscal system is through the concept of AGR.

AGR

An important and relatively new concept is that of AGR. Under an R/T system with no formal cost recovery limit, companies normally recover costs out of net revenue. AGR is, therefore, limited only by the royalty. If there are sufficient eligible deductions, the company would pay no tax in a given accounting period. In most R/T systems, there is no limit to the amount of deductions that a company can take in any given accounting period, and a non-tax-paying position is a possibility under these regimes.

Under a production-sharing system with a 60% cost recovery limit and taxes paid by the NOC out of its share of profit oil, the contractor can recover costs of up to 60% of gross revenues in a given accounting period. Unrecovered costs are ordinarily carried forward. However, if the contractor has a 45% take, then the contractor gets 45% of the remaining 40% (of revenues) or another 18%. With sufficient deductions (or cost recovery requirements) the contractor can access up to 78% of gross revenues (or up to 78% of his working interest share of gross revenues). This is very close to the world average AGR of 81%. (*See* Tables 2–3 and 2–4.)

Table 2–3 Access to Gross Revenues Calculation PSC Structures

Take Calculation		Access to Revenues Calculation		
$100.0 -10.0	MM	$100.0 -10.0	MM	Gross revenues Royalty
90.0 -30.0		90.0 -60.0*		Net revenues Cost recovery (limit = 60%)
60.0 -36.0		30.0 -18.0		Profit oil 60% to government
24.0		12.0		40% to contractor
-7.2		0		**Corporate income tax 30%**
16.8		12.0		Contractor net income after tax
24.0%				**Contractor take:** Contractor net income after tax ÷ (gross revenues – costs)
		72%		Access to gross revenues: Cost oil + after-tax profit oil (60 + 12)

* Costs must equal full cost recovery limit for AGR calculation.

Table 2–4 Access to Gross Revenues Examples

Country	Type of System	Royalty	Cost Recovery Limit	Access to Gross Revenues
U.K.	R/T	0%	100%	100%
New Zealand	R/T	5%[1]	100%	95%
Philippines	S/A	-7.5%[2]	70%	87%
World Average	N/A	7%	80%	81%
Indonesia (std)	PSC	0%	80%	86%
Malaysia (std)	PSC	10.5%[3]	50%	68%

[1] 5% ad valorem royalty (AVR).

[2] Filipino participation incentive allowance (FPIA), part of service fee paid to contractor group as function of gross revenues if qualified.

[3] 10% royalty + 0.5% research cess (effectively a severance tax based on gross revenues).

R/T: royalty tax system
S/A: service agreement
PSC: production-sharing contract
N/A: not applicable
std: standard

Both royalties and cost recovery limits guarantee that in any given accounting period, the government will get a share of gross revenues or production—regardless of whether true economic profits have been generated. At the end of the life of an oilfield, the government-guaranteed share of gross revenues takes on all of the characteristics of a royalty, and most fiscal systems need re-thinking in this respect.

GOVERNMENT PARTICIPATION / CARRY

Many systems provide an option for the NOC to participate in development projects. Under most government participation arrangements, the contractor bears the cost and risk of exploration and, if there is a discovery, the government backs-in for a percentage. In other words, the government is *carried* through exploration. This is fairly common and automatically assumed whenever some percentage of government participation is quoted. Both the Indonesian and Malaysian PSCs have government participation clauses, but Indonesia rarely exercises its option to participate.

The financial effect of a government partner is similar to that of any working interest partner with a few major exceptions. First, the government is usually carried through the exploration phase and may or may not reimburse the contractor for past exploration costs. Second, the government contribution to capital and operating costs is normally paid out of production. Finally, the government is seldom a silent partner.

Approximately 40% of the countries in the world have the option to back-in at the point of commerciality. Of these, the percentage share of participation ranges from 10% to more than 50%. The average carry for those governments that have the option is 30%. Factoring in the 60% of the countries that do not have the option to participate, the world average is 13%.

RINGFENCING

The issue of recovery or deductibility of costs is further defined by the revenue base from which costs can be deducted. Ordinarily all costs associated with a given block or license must be recovered from revenues generated within that block. The block is *ringfenced*. This element of a system can have a huge impact on the recovery of costs of exploration and development. Indonesia requires each contract to be administered by a separate new company. This restricts *consolidation* or effectively erects a ringfence around each license area.

Some countries will allow certain classes of costs associated with a given field or license to be recovered from revenues from another field or license. India allows exploration costs from one area to be recovered out of revenues from another, but development costs must be recovered from the license in which those costs were incurred.

From the government perspective, any consolidation or allowance for costs to cross a ringfence means that the government may in effect subsidize unsuccessful exploration efforts. This is not a popular direction for governments because of the risky nature of exploration. However, to allow exploration costs to *cross the fence* can be a strong financial incentive for the industry.

If a country with an effective tax burden of 60% allowed exploration costs to be deducted across license boundaries (consolidation), then the industry would be effectively drilling with "dollars" worth only 40 cents. It would more than cut the risk in half. From the perspective of the development engineer, it has little meaning unless development and operating costs are also allowed to cross the fence. Dropping or loosening the ringfence can provide strong incentives, especially to companies that have existing production and are paying taxes.

About 80% of the countries in the world have a ringfence of some sort. Sometimes there is a limited ringfence around contiguous blocks or around upstream activities, or the ringfence may apply to cost recovery but not corporate income taxes. Sixty-four percent (64%) of the countries have a typical ringfence that simply isolates each license area.

CRYPTO AND OTHER TAXES

On most occasions, comparison of one system to another ignores the effects of ancillary fiscal elements such as bonuses, rentals, training fees, "social" obligations, severe procurement restrictions, etc. For example, the 1996 bid round in Venezuela set the division of profits for most of the blocks around 90/10% to 88/12% in favor of the government (ignoring the 35% government carry). However, on two of the blocks, the signature bonus exceeded US $100 MM. The effective division of profits for the blocks will be lower for the contractor groups who paid the big bonuses, but it is hard to say what that might be without indexing the calculation to an assumed prospect or field size.

WORLD AVERAGE FISCAL SYSTEM

For the purpose of providing a frame of reference, a world average fiscal system is summarized in Table 2–5. It is contrasted with world average PSC and R/T systems. Table 2–6 illustrates the kind of peer group analysis that countries are doing regionally and globally.

Table 2–5 World Average Fiscal System Statistics

	Production-Sharing Contracts	World Average System	Royalty Tax Systems
Number of systems	**68**	**123**	**55**
Contractor take	34%	38%	42%
Royalty	5.7%	7.1%	8.9%
Cost recovery limit	63%	79%	98%
Access to gross revenues	73%	81%	90%
Government carry (40% have a carry)	12%	13%	15%

Table 2–6 Regional Comparison—Southeast Asian Frontier

	Malaysia Frontier	Eastern Indonesia	Timor Gap	The Philippines	Australia Federal	New Zealand
Type of system	PSC	PSC	PSC	SA	R/T	R/T
Contractor take	32%	30%	26%	46%	46%	56%
Royalty	10.5%	3%[2]	0%	-7.5%[3]	0%	5%
Cost rec. limit	75%	80%	90%	70%	100%	100%
Access to gross revenues	87%	91%	93%	87%	100%	95%
Govt. carry	15%	0%	0%	0%	0%	0%
Ringfence	Yes [1]	Yes	Yes	Yes	No[4]	Yes

[1] Everything is effectively ringfenced but contiguous blocks may qualify for consolidation.

[2] The Domestic Market Obligation is effectively a royalty with a 5-year holiday. The effective rate is approximately 3%.

[3] The Filipino Participation Incentive Allowance (FPIA) grants as part of the service fee 7.5% of gross revenues to the contractor group with sufficient Filipino ownership. Effectively a negative royalty!

[4] No ringfence for exploration costs *offshore*. Fields are ringfenced for development costs. No ringfence for corporate income taxes.

COMMENTARY

Financial analysis of a fiscal system focuses on the division of revenues, division of profits, and limitations on the recovery of costs. Determining how profits are divided drives a wedge of clarity right to the heart of any fiscal system or contractual arrangement. Even though royalties are captured in the *take* statistics, the royalty rate is a barometer of system efficiency. Additional perspective is gained by evaluating AGR, which illustrates how quickly costs can be recovered, as well as demonstrating the relative efficiency of a system. Cost recovery limits commonly found in PSCs are only part of the picture; AGR adds the other important dimension. Crypto taxes are almost by definition unquantifiable and certainly difficult to collapse into the take statistic, but they can have a dramatic influence in some cases. In the end, analysis of a given fiscal system is part science and part art—as perhaps it should be.

What has changed since 1996?

Government Take vs. State Take

In the past, there were a number of analysts who used different terms (such as government take or state take) to indicate whether or not the government participation element had been factored-in to the *take* calculation. In this chapter, *government take* did not include this element while *state take* did. This kind of distinction is rare as of this writing (2003), because in many languages interpreters see no difference between the words *government* and *state*. Whether or not the government participation option should be factored-in has been a source of debate in the industry. It appears to have been resolved. Now almost everybody agrees that it should be factored-in when comparing contracts or systems for exploration rights. Since writing this for *PAFMJ* (early 1996), I have abandoned the terminology (*state* take vs. *government* take). It is too cumbersome and confusing. Now I use the term *government take*, which includes all means by which the state (or government) obtains a piece of the pie. There are situations where the government participation element must be excluded and treated separately and this is discussed later in this book.

AGR

My article in 1996 introduced the concept of AGR, which is the complement of ERR. Since then, the term *ERR* is more commonly used. Either statistic provides added insight into the government take statistic—going beyond *what* the percentage take is and showing *how* the government obtains its share of the pie.

Other metrics

The state-of-the-art has improved. Since 1996, two other important statistics have been added: the *savings index* and the *entitlement index*. These are discussed further in other chapters of this book.

REFERENCES

Johnston, D. *International Petroleum Fiscal Systems and Production-Sharing Contracts*, Tulsa, Oklahoma: PennWell Publishing Company, 1994.

Johnston, D. *International Petroleum Fiscal Systems and Production-Sharing Contracts Course Workbook*, Daniel Johnston & Co., Inc., 2002.

Johnston, D. "Global Petroleum Fiscal Systems Compared by Contractor Take," *Oil & Gas Journal*, 12 December, 1994 pp. 47–50.

Johnston, D. "Different Fiscal Systems Complicate Reserve Values," *Oil & Gas Journal*, May 29, 1995, pp. 39–42.

Van Meurs, P. *World Fiscal Systems for Oil*, New York: Barrows Inc., 1993.

Van Meurs, P. *Acreage Laws & Tax—Annual Review of Petroleum Fiscal Systems*, London: Petroconsultants, 1995.

3

Thinking of Going International?
Some Useful Tips

Engineering, geological, and financial principles are relatively universal throughout the petroleum industry, but the difference in scale between ordinary U.S. operations and overseas projects can be staggering. Many provinces, geological basins, and plays overseas are at a stage now that existed in the United States two to three generations ago.

More than 3.3 million wells have been drilled in the United States. Nothing anywhere else compares. The chances of making a large discovery overseas are about 30 times better than in the United States. This can be seen in Table 3–1.

Table 3–1   New Field Wildcats per Discovery

	International			United States		
Discovery (MMBLS)	>50	>100	>500	>50	>100	>500
Discoveries per wildcat	1/26	1/40	1/165	1/690	1/1270	1/5732

Source: Marlan Downey (Dallas Energy Symposium Speech, March 1995)

Production rates in the international sector are also more promising than in the United States. For example, 75% of the nearly 600,000 producing oil wells in the United States are classified as *stripper wells*, and the average production rate of these 450,000-odd stripper wells is about 2 BOPD.

Table 3–2 summarizes typical production rates from around the world. Table 3–3 illustrates test rates for 1996 discoveries (*see* also Table 3–4). Test rates quoted in the press and industry literature are often *combined* flow rates from more than one zone and are seldom representative of the kind of production that might be expected from a typical well for the first full year of production—the information needed for economic analysis. Drillstem or production tests provide a wealth of information regarding petrophysical characteristics of the reservoir, reservoir pressures, reservoir boundaries, fluid properties of viscosity, specific density, API gravity, etc. But making projections of well deliverability from well test rates must be done with caution. The average daily production rate for a typical development well during the first full year of production can often be five times less than quoted test rates from the discovery well.

Table 3–2 Comparative Oil Well Production Rates

	Average 1st year (BOPD)	1997 Average (BOPD)
Lower 48 (onshore)	30–100+	2–30
Gulf of Mexico	300–400	180–200
Alaska	1,500–2,000+	1,000
U.K.	8,000–10,000	3,000
Vietnam	7,000–8,000	6,000
Angola	3,000	1,600
New Zealand *	1,500	1,000
Maui B	**	10,000
Tunisia	1,000	500
Malaysia	2,000–3,000	1,300
Kuwait	5,000–10,000++	2,500

 * Offshore New Zealand
 ** 1997 is first year of oil production at Maui B.

Table 3–3 International Discoveries—1996

			Discovery Well Test Rates		
Category	#	Bottom 25%	Average	Top 25%	
Oil	33	940	4,600	9,900	BOPD
Gas	43	42	8	58	MMCFD
			1,300		BCPD
			47		BBLS/
					MMCF
Oil & gas	27	840	3,700	7,800	BOPD
		3	11	33	MMCFD
Total discoveries	103				
Onshore	53				
Offshore	50				

Major discoveries compiled by Petroconsultants. Source: AAPG *Explorer*, January 1997.

Table 3–4 Worldwide Production Statistics (1995)

Country	Producing Oil Wells	Average Daily Production (BOPD)	Average BOPD per Well
Abu Dhabi	993	1,818,000	1,830
Algeria	1,273	760,000	597
Angola	403	640,000	1,588
Argentina	11,709	714,000	61
Australia	1,209	558,000	461
Austria	1,145	23,000	20
Bahrain	376	105,000	280
Bangladesh	31	1,100	35
Barbados	90	1,300	14
Benin	6	2,500	417
Bolivia	384	28,000	73
Brazil	6,673	696,000	104
Brunei	858	159,000	185
Cameroon	153	100,000	653
Canada	43,758	1,798,000	41
Chile	390	10,500	27
China	49,700	2,989,000	60
Colombia	5,819	580,000	100
Congo	409	178,000	435
Croatia	901	37,000	41
Denmark	151	186,000	1,232
Dubai	151	335,000	2,219
Ecuador	1,021	392,000	384
Egypt	1,228	890,000	724
France	465	50,000	107
FSU	122,820	6,950,000	57

Table 3–4 Worldwide Production Statistics (1995) (continued)

Country	Producing Oil Wells	Average Daily Production (BOPD)	Average BOPD per Well
Gabon	379	354,000	934
Germany	1,605	57,000	36
Ghana	3	6,000	2,000
Greece	12	9,400	783
Guatemala	14	9,800	700
Hungary	1,138	70,000	62
India	3,344	708,000	212
Indonesia	8,622	1,329,000	154
Iran	751	3,654,000	4,866
Iraq	58	600,000	10,435
Israel	9	100	11
Italy	219	92,000	420
Ivory Coast	0?	10,000	
Japan	223	15,200	68
Jordan	4	60	15
Kuwait	700	1,800,000	2,571
Libya	1,087	1,370,000	1,260
Malaysia	550	685,000	1,245
Mexico	4,740	2,689,000	567
Morocco	9	200	22
Myanmar	450	16,000	36
Netherlands	208	67,500	325
Neutral Zone	614	400,000	2,158
New Zealand	62	29,000	468
Nigeria	1,938	1,887,000	974
Norway	436	2,720,000	6,239
Oman	2,099	846,000	403
Pakistan	120	55,000	458
Papua New Guinea	29	100,000	3,448

Table 3–4 Worldwide Production Statistics (1995) (continued)

Country	Producing Oil Wells	Average Daily Production (BOPD)	Average BOPD per Well
Peru	3,568	129,700	36
Philippines	8	3,300	413
Poland	2,179	8,000	4
Qatar	288	438,000	1,521
Ras Al Khaima	7	1,000	143
Saudi Arabia	1,400	7,867,000	5,626
Serbia	646	20,000	31
Sharjah	48	50,000	1,042
South Africa	9	10,400	1,156
Spain	36	15,000	417
Suriname	317	7,000	22
Syria	964	606,000	629
Taiwan	85	1,000	12
Thailand	392	52,000	133
Trinidad & Tobago	3,346	132,700	40
Tunisia	187	88,000	471
Turkey	776	68,000	88
UK	839	2,515,000	2,998
USA	582,768	6,545,000	11
Venezuela	14,789	2,565,000	173
Vietnam	28	176,000	6,285
Yemen	243	345,000	1,420
Zaire	128	28,000	219
My count	894,590	61,251,760	68
O&GJ	903,770	61,444,800	68

Source: Oil & Gas Journal, Worldwide Production Report, Decmeber 1995.

FISCAL TERMS

As illustrated in Figure 2–1, fiscal terms in the international sector are varied and complex. The *X*-axis of this graph depicts the *state take*—the government share of profits (undiscounted) over the life of a field or license. The state take statistic employed here is consistent with the Petroconsultants' (Geneva—now IHS Energy) usage of the term, which includes the impact of government participation/risk free carry. Terminology regarding division of profits is diverse, non-standard, and slightly controversial; investors are well advised to insist on precise definition of all terms.

State take (also known as government take) ranges from as little as 25% (Ireland) to more than 90% (Venezuela). In many of the countries where geological prospectivity is fairly good, the government share of profits will typically range from 60 to 80%—quite a wide range with lots of room for negotiation. The difference between a state take of 75% and a state take of 80%, for example, could result from any one of the following:

- An additional 13–15% royalty

- An additional 20% profits-based tax

- An export tariff of $\pm$\$2.50/BBL (depending on wellhead price)

- An additional 20% government carry

- A large bonus, depending on degree of risk

GETTING STARTED IN THE INTERNATIONAL OIL AND GAS BUSINESS

Oil and gas businesses typically enter the international market through three different routes:

- Grassroots exploration
- Farming in
- Acquisitions

Grassroots Exploration

Grassroots exploration has a lot of appeal because it normally entails a lower entry cost. However, by pursuing this method, a relatively small company will have difficulties obtaining a respectable international portfolio very quickly. Some countries can easily tie up several years of effort before signing a single contract.

The typical strategy is to obtain an exploration license in an area with high prospective potential, farm out the costly work commitment, and get carried through the first well or two. This, however, is getting more and more difficult to do. From the government perspective this is referred to as *license trading*. While many governments do not like such an arrangement, other governments understand the marketplace and know that finding partners is a normal part of the business. As competition has escalated over the years, governments have become wary of explorers who do not have sufficient capital or money-raising capability to fulfill work programs.

Farm-in Strategy

Numerous interesting opportunities are available in the current buyers' market. A company with sufficient funds to actually participate in a few wells could conceivably sift through hundreds of projects.

A difficult hurdle for many companies is the concept of being *promoted*, and this is unfortunate because there should be no stigma attached to farming into a project unless, of course, the company doing the farming-in does not understand market conditions and overpays.

The time-honored *third-for-a-quarter promote* is still a common formula in the industry. Under this arrangement, companies farm into an exploration license by paying all of the costs of the first well or so for 75% of the working interest. Thus, the grassroots explorationists who obtained the license would be carried for 25% of the working interest. In areas where prospects are particularly lucrative, the carry may amount to 50%. By the late 1990s, the amount of *promote* or carried working interest a company farming-out could expect had dropped. Many deals had carried interests as low as 15% or less with some deals referred to as *ground floor* where there is no *promote*.

In today's fairly competitive and efficient market, few companies are willing to pay much up-front cash to the lease-holder. Sometimes the explorationists are able to get reimbursement on a portion of past costs, typically geological and geophysical work (G&G), as part of the deal.

Acquisitions

One fast-track approach to establishing a presence in a region is to acquire an interest in a license or a producing property. The ideal acquisition would include delineation and exploration upside. Once sufficient experience is gained in the area from the acquired working interest, then the company may branch out and acquire additional interests either through grassroots exploration efforts or by farming in.

Unfortunately, in many oil and gas production acquisitions, the buyer pays too much. This holds true for numerous petroleum acquisitions and is not peculiar to international transactions. But competition provides the dynamic, and competition is greater now than ever before.

Problem Areas

There are certainly a number of rewards associated with the international scene, but there are also plenty of pitfalls. The savvy businessperson will avoid falling prey to the following:

- Underestimating time required to negotiate anything

- Underestimating costs

- Overestimating wellhead price for crude oil and gas

- Overestimating the influence of local agent or contact

- Contract terms that contradict petroleum legislation

- Getting a start in the international business in the FSU

HOT SPOTS WORLDWIDE

While there are certainly some interesting hot spots, companies with limited capital should pursue a regional focus. Even large companies have developed a regional focus strategy but spread across a number of regions.

Argentina Good fiscal terms, good business climate, exploration potential like some of United States 30 years ago. Typical exploration well cost: US $3–4 MM.

West Africa Rich/virgin exploration potential, fair to good terms, lots of deepwater, harsh political risks. Typical exploration well cost: shallow water US $8–12 MM; deepwater $14 -35 MM.

United Kingdom Excellent terms, excellent infrastructure, market for gas, very good business climate,

	Greenpeace risk, geological targets smaller. Typical exploration well cost: US $7–9 MM.
Australia	Very good fiscal terms, good business climate, variety of plays onshore/offshore, market for gas in many areas, relatively low cost environment.
Gulf of Mexico	Excellent terms, no ringfencing, market for gas, excellent infrastructure, multi-pay prospects but smaller targets, low-cost environment except for deepwater. Typical exploration well cost: US $3–4 MM; deepwater $20–30++ MM.
China	Terms are negotiable but government understands market, all kinds of deals available onshore/offshore, high risk/low risk, etc. Bohai Bay, shallow water, low-cost proven hydrocarbon province.

FROM THE GOVERNMENT'S PERSPECTIVE

Most governments want legitimate companies to come in and drill exploratory wells or participate in some variation on this scenario. While some government representatives carry themselves with confidence and a sense of purpose, others even at high levels, quiver with anxiety over four fundamental issues that seem to plague every government. Successful companies will address these concerns.

Type of System—
Division of Profits and Revenue Protection

Signing a bad deal is a basic human fear—the fear that the other party has negotiated a windfall. In the high-pressure environment in which government representatives work, their anxiety is easily understandable. Government officials do their negotiating on a stage that is scrutinized by other governmental agencies; the ruler, dictator, or prime minister; the citizens; and the press.

Cost Control

Some governments harbor a morbid fear that oil companies will raid the governmental treasury through frivolous spending, as well as cheating on costs, i.e., inflating spending claims. In actuality, governments hold most (but not all) of the cards on this issue. Their primary means of protection include:

- The budget process and authorizations for expenditure (AFEs)

- Procurement process and thresholds

- Audits (here accountants are worth their weight in gold!)

- Laws and penalties for non-compliance or fraud

- Natural tendency of oil companies to economize

- Non-operator partners or watchdogs

Maximum Efficient Rate

The conventional wisdom in almost all governments (aside from the odd engineer in the NOC) is that the oil company's only concern is short-term profits: that they will produce oil too quickly and inefficiently and leave behind more than they otherwise would.

Evaluating the Companies

Governments have a hard time understanding the numerous companies, business combinations, entrepreneurs, their financial wherewithal, and their true agenda. Government personnel would like to be able to determine the true credit-worthiness/financial strength, past business practices and reputation, technical ability, and international experience of the various companies who come calling. But it is difficult for them to do this. The information is not easy to obtain. Usually they only have what the companies provide for them.

WHAT ABOUT THE FSU?

For most of the early 1990s, the FSU experienced a black gold rush of biblical proportions. Plane loads of *entrepreneurs* arrived like locusts. But after more than a decade of activity, is anybody making money producing oil there? I don't think so. Most projects now are dead in the water. By the end of the decade, the industry's expectations compared to their experiences are like the difference between good caviar and bad caviar—a huge difference.

The FSU opened its doors at a time when many major oil companies were realizing that their exploration efforts had not been so successful. Many companies have redirected capital toward less risky ventures, such as large-scale development or rehabilitation projects, which the FSU has in abundance.

Many projects have stalled because firm agreements have not been hammered out. One huge obstacle for many was the insistence of the Russians to impose the famous *export tariffs*, which typically ranged from around $4.50 to $6 per barrel. These kinds of tariffs could kill nearly any deal, yet they were very intransigent about the tariffs. And one of the biggest stumbling blocks was the issue of ownership.

Ownership

The culture clash that Western companies encounter when doing business in the FSU is dramatic compared to many other countries and its critical focus is the issue of ownership. The Russians are saddled with boundary conditions and established bureaucracies that provide nearly insurmountable obstacles to moving forward. The FSU went from serfdom to communism in the wink of an eye. Western concepts of ownership and profit, while understood by some in the FSU, have simply not been internalized or generally accepted. Beyond that is the more important issue of the distribution of wealth. Who is going to get what?

There is a struggle between authorities at the federal level, the Oblast level, the municipal level (weak as that link is), the geophysical associations (who happened to have found the oil), and the production associations who are now sitting on it. Nobody wants to get left out, and who can blame them?

The issue of ownership is further complicated by the JV nature of most of the projects. The production associations who are in physical possession of the oilfields are populated with (unpaid) workers whose families have worked in the oilfields for three and four generations.

Chemistry

As with any business arrangement, good *chemistry* between partners is crucial, yet impossible to stipulate in the language of a contract. Goodwill can be a magical thing, and many of the business discussions and JVs have started out with a great deal of that. However, the disparity of incomes between workers and professionals in the FSU and the West is a great source of problems. Many foreigners make in one hour what their Russian counterpart makes in a month—if and when he gets paid.

The cultural gap is also substantial. Often when Westerners are convinced that an *agreement* has been reached, their counterparts are just warming up to further negotiations. The typical Western concepts of timing and urgency do not compute.

Division of Profits

The cornerstone of any business relationship is the division of profits. Even in the Western world, there are diverse perspectives on this issue. There is little experience in the FSU regarding the world market for oil deals. As a result, this is one area of critical discomfort and insecurity for most of the former Soviet governments. Some of the existing contracts are ripe for renegotiation, some were born dead, and numerous others represent nothing more than a virtual right to negotiate further.

Petroleum Law

One of the most common complaints is the lack of legal precedence. This is a legitimate complaint considering that few of the established laws deal with petroleum exploration. Furthermore, the laws do not clearly grant any particular organization sufficient authority to negotiate and close deals with any degree of independence or confidence.

Nagging concerns exist over inconsistencies in Western logic. For example, outsiders negotiating in the FSU insist on binding international arbitration clauses in the contracts for dispute resolution. The question inevitably arises as to how disputes are resolved in America. The answer, of course, is that the US courts resolve many of these issues as litigation is pursued. The irony is not lost on the Russians.

The solution to the problem is not as simple as getting PSC legislation through the Duma. The infrastructure for capturing and disbursing the equivalent of the myriad taxes, levies, imposts, duties, royalties, tariffs, and excises that now exist must be rationalized. This is no easy task.

COMMENTARY
DOOM AND GLOOM?

Many companies have already overextended themselves and have little to show for it. Perhaps they should have waited. Russia is not the place for a company to get its start in the international oil business. Companies with extensive international experience can adopt a business-as-usual attitude with the understanding that it is just a matter of more time and money. The risk is that they may someday wish they had spent it somewhere else. But it is a risk they are willing and able to take. The oil is there.

4

Trends and Issues in Foreign PSCs

The international petroleum industry provides substantial opportunities and challenges, and it is possible that the challenges will always outweigh the opportunities. While the U.S. oil patch has had much to recommend it in the past, it now has formidable competition overseas. Furthermore, some countries are more user-friendly and provide a more hospitable business environment than what is found in some parts of the United States.

The key differences in the international oil patch are:

• Geopotential and costs

• Negotiations and fiscal terms

GEOPOTENTIAL

While geological and engineering principals are relatively universal, exploration potential is dramatically better outside most of the United States because of the advanced maturity of the U.S. oil patch. Most foreigners are shocked to hear U.S. production statistics. For example, there are nearly 600,000 oil wells in the United States that produce an average of 11–12 BOPD. If the 1500-odd Alaskan wells (around 1000 BOPD each) are excluded, the average lower 48 oil well produces less than 10 BOPD.

The United States is unique with its stripper well category for oil wells—those that produce an average of less than 10 BOPD. Nearly 75% of the oil wells in the United States are stripper wells. Of these 450,000-odd wells, the average production rate is around 2 BOPD.

It is a function of maturity. Our industry has drilled more than 3.3 million wells here in the United States. By comparison, in the vast provinces of the FSU, less than a million wells have been drilled.

Some of the hot spots worldwide are summarized in Table 4–1. The cost of an exploration well provides a useful index for both risk capital as well as potential development costs. Drilling costs can often represent from one-third to one-half of development capital. Table 4–2 shows reported well test rates from reported 186 discoveries in 1996–97. Well test rates will be higher than the average production rate per well during the first full year of production. Yet, the test rates are a very useful indicator.

Table 4–1   Selected Hot Spots Worldwide

	1996–1997 Discovery Well Test Rates		Typical Exploration Well Cost	Typical State Take
	Oil (BOPD)	**Gas (MMCFD)**	**($MM)**	**(%)**
United Kingdom	8,000	22	$8–10	33%
Asia				
W. Indonesia	560+	23	$3–4	86%
	560+	23	$3–4	70%
Vietnam	N/A	N/A	$8–12+	57–88%
Australia				
Offshore	7,600 max	10	$3–6+	61%
Pakistan				
Onshore			$2–4+	50–60%
Offshore				38–44%
Latin America				
Venezuela	4,000+	10+	$3–16	90+%
Argentina	750	17	$1.5–4	40+%
W. Colombia	700		$0.3–0.5	80+%
E. Colombia	3.500+		$20–30	80+%
Peru				60+%
Africa				
North Africa	4,000	30	$2.5–6	Various
West Africa	6,000		$7–15+	Various
Gulf of Mexico				
Deepwater	800+	15+	$5–13+	40%+
Shallower water			$3–4	43–48%

Table 4–2   Reported Well Test Rates from 186 Discoveries Worldwide 1996–97*

	Lower Quartile	Average	Upper Quartile
BOPD	840	5,700	10,400
MMCFD	4	23	50

* Many of the test rates are combined flow rates from multiples zones.

Fiscal Terms

Analysis of the commercial terms of any fiscal system or contract requires detailed cash flow modeling and analysis as well as a bit of art. When the exercise can distill the analysis into the following elements, which comprise the key commercial aspects of a fiscal system, comparison of various contracts/systems is fairly efficient:

- Signature bonus*†
- Contractor take
- Government participation (typically a "carry")
- Royalty†
- Cost recovery limit†
- AGR/revenue protection (RP) [RP is the equivalent of an ERR.]
- Entitlement index
- Ringfencing
- Crypto taxes‡ (see examples)
- Government Grief Index (GGI)‡

* While some of these statistics are embodied in the *take* and RP statistics, they still provide additional insight.

† While bonuses are captured in many *take* statistics, they never really get full credit.

‡ Qualifiable—not quantifiable—in my opinion this stuff is nearly pure gut feel. The term *GGI* is borrowed from Richard Barry's book *The Management of International Oil Operations*. It is quite a useful and self-explanatory term.

The hard part is to combine these elements into a meaningful analysis within the context of the geopotential coupled with the cost of doing business and GGI (political risk is part of this, of course). It is a challenge, and the marketplace is more dynamic than ever.

STRENGTHS AND WEAKNESSES OF TAKE STATISTICS

The division of profits provides substantial insight into the relative virtues of a contract or fiscal system. The take statistics (which deal with the division of profits) provide most of the answers to the question "How tough are these terms?"

EXAMPLE CALCULATION— DIVISION OF PROFITS

Take calculation

Full cycle

$100.0 MM	Gross revenues
-10.0	Royalty
90.0	Net revenues
-30.0	Cost recovery
60.0	Profit oil
-36.0	60% to government
24.0	40% to contractor
-7.2	Corporate income tax 30%
16.8	Contractor net income after tax

24.0% Contractor take

Contractor net income after tax

$\div$ gross revenues - costs

$[16.8 \div (100-30)]$

76.0% Government take

$[10 + 36 + 7.2 \div (100-30)]$

However, there are some pitfalls. For example bonuses often do not really get appropriate representation in the take statistics. As far as government take is concerned, a one-dollar ($1.00) bonus can represent an infinite take for a block or license in which no discovery is made. But if a discovery is made or contemplated, the bonus can practically disappear. In the aggressively attended licensing round in Venezuela in 1996, the La Cieba block received a US $104 MM bonus from the Mobil-led consortium. Part of the reason for the large bonus was that there was a discovery well on the block that had tested around 4000 BOPD (combined flow rates). If the potential of this discovery was estimated at, say, 500 MMBBLS of recoverable oil, then revenues over the life of the field could be on the order of $7.5 billion.

With total expected profits of around $5 billion, the $104 MM bonus suddenly begins to disappear. It represents only 2% of profits. The famous Venezuela take statistic hardly wiggles:

VENEZUELAN STATE TAKE

Without $104 MM bonus 92.2%
With the $104 MM bonus 92.5%

DIVERSITY OF TERMINOLOGY

The terminology in the industry regarding the division of profits is diverse and confusing. Much of the confusion centers around how government participation is treated. For many countries or contracts, two statistics are used, each of which requires different terms. The list that follows explains some of the diverse terminology. Notice the additions to this list from the previous chapter.

THE DIVISION OF PROFITS

"The Split"

Government	Company/contractor
• Government equity split	• Contractor equity split
• Government after-tax equity split	• Contractor after-tax equity split

Almost obsolete terminology consistent only (technically) with Indonesian PSC.

• Government take (excludes government participation)	• Contractor take
• State take (includes government participation)	• Contractor take

Petroconsultants (Geneva) use this terminology.

• Government marginal take	• Contractor marginal take
• Net government take on the marginal barrel	• Net contractor take on the marginal barrel

This terminology is used mostly in U.K.

• Tax take (= Petroconsultants' government take)	• Contractor take
• Government take (= Petroconsultants' state take)	• Contractor take

This is the terminology used at the University of Aberdeen. Pedro Van Meurs / Barrows use government take in this context.

• Discounted government take	• Discounted contractor take

The picture changes dramatically from present value point of view (in favor of government); nevertheless, the take statistics are very valuable and useful.

• Government profit share	• Contractor profit share

- • Net cash margin
- • Fiscal take
- • Fiscal net
- • Bottom-line financial split/bottom-line income split
- • Net net (this is a landman term)

No such thing as "no bonus"

Some countries are proud to point out that they require no bonus. This certainly adds an appealing element for countries that must compete for capital and technology. However, most of the countries that have no bonuses allocate exploration rights on the basis of *work program*. Work program bidding though can have many of the characteristics of bonus bidding. Instead of *money left on the table*, we find *wells left on the table*. And the concept of *the winner's curse* works here as well. The winner's curse is the difference between the highest bid and the next highest bid. Some view it as the difference between the highest bid and the average bid.

ERRs/RP

RP or the ERR is the minimum share of revenues the government will get in any given accounting period (excluding any working interest share if the NOC participates). In any given accounting period, there is a limit to just how great a share of gross revenues a company may have access to. The complement of AGR is either the royalty rate or the ERR. For example, government RP is 14% in the Indonesian standard oil contract; thus, an oil company's AGR is 86%. AGR is an index that measures the maximum share of gross revenues a company may receive in any given accounting period. Assuming the oil company had a 40% working interest, the maximum share of revenues in any given accounting period would be 34.4% (86% of 40%). The entitlement index often represents the percentage of proved recoverable reserves that may be "booked" according to standard SEC criteria. Example calculations are shown on as follows.

With R/T systems, the upper limit is usually due to the royalty alone. Therefore, AGR with R/T systems is net revenue, and legal entitlement to the hydrocarbons is often effectively the same. With

PSCs, the calculation is different because of the cost recovery limit. Cost recovery limits can create much the same effect as a royalty. The AGR statistic combines the royalty effect created by a cost recovery limit with the royalty. Assuming there are sufficient eligible deductions, the company can obtain up to the full cost recovery limit and then the resulting pre-tax share of profit oil.

AGR CALCULATION

Single accounting period

$100.0	M	Gross revenues
-10.0	M	Royalty
90.0		Net revenues
-60.0*		Cost recovery (limit = 60%)
30.0		Profit oil
-18.0		60% to government
12.0		40% to contractor
-0		Corporate income tax 30%
12.0		Contractor net income after tax
72.0%		AGR (cost oil + profit oil) (60 + 12.0)
28.0%		RP or ERR (Royalty + government profit oil) (10 + 18.0)

* Costs must equal full cost recovery limit for AGR calculation (assume unlimited deductions).

Emphasis is placed these days on the reserves a company can *book*. The criteria for reserves recognition (*booking*) is summarized in Table 4–3. For a PSC, the entitlement is based on two components: cost oil and profit oil. An example estimate of reserves entitlement is the entitlement index calculation where 54% of recoverable reserves could be *booked* or added to reserves of the company.

Table 4–3 SEC Reserves Recognition—Key Criteria

• Right to extract	• Right to take in kind (not critical)
• Actual transfer of title, wellhead—export point (not critical)	• Elements of risk and reward
	• Economic interest

Governing Definition for Reserves Recognition
SEC Regulation SX Rule 4-10

(1) Mineral interests in properties:

(i) fee ownership or a lease, concession or other interest representing the right to extract oil or gas subject to such terms as may be imposed by the conveyance of that interest

(ii) royalty interests, production payments payable in oil or gas, and other non-operating interests by others

(iii) those agreements with foreign governments or authorities under which a reporting entity participates in the operation of the related properties or otherwise serves as *producer* of the underlying reserves (as opposed to being an independent purchaser, broker dealer, or importer).

Properties do not include other supply agreements or contracts that represent the right to purchase, rather than extract, oil and gas.

From: Claude C. McMichael and Elliot D. Young, 1997 SPE Hydrocarbon Economics and Evaluations Symposium, Dallas, Texas, 16–18 March, 1997. SPE Paper 37959: "Effect of Production Sharing and Service Contracts on Reserves Reporting"

ENTITLEMENT INDEX CALCULATION

Full Cycle

$100.0	M	Gross revenues
-10.0	M	Royalty
90.0		Net revenues
-30.0		Cost recovery
60.0		Profit oil
-36.0		60% to government
24.0		40% to contractor
-7.2		Corporate income tax 30%
16.8		Contractor net income after tax
54.0%		**Contractor entitlement**

Cost oil + profit oil (30 + 24)

Table 4–4 shows the entitlement index for PSCs compared to royalty tax systems and the world average. Also reflected are some comparative statistics on key elements in evaluating world fiscal systems. Table 4–5 depicts some typical key contract elements currently effective.

Table 4–4 World Fiscal System Statistics

	PSCs	World Average System	R/T
Number of systems	68	123	55
Government take	70%	67%	64%
Government participation (40% have a carry)	12%	13%	15%
Royalty	5.7%	7.1%	8.9%
Cost recovery limit	63%	79%	98%
Access to gross revenues (AGR)	73%	81%	90%
Effective royalty rate ERR	27%	19%	10%
Entitlement index	69%	79%	91%

	The More Prospective Countries		
	PSCs	Average	R/T
Number of systems	28	44	16
Government take	75%	73%	72%
Government participation	13%	17%	25%
Royalty	6.1%	7.7%	10.5%
Cost recovery limit	62%	75%	98%
Access to gross revenues (AGR)	71%	77%	88%
Effective royalty rate (ERR)	29%	23%	12%
Entitlement index	65%	73%	88%

* The government take statistic includes the government participation

Table 4–5 Typical Ranges of Key Contract Elements

Area: Wide variation   Average **150,000-350,000 acres ±**

Duration: Exploration   Multi-phase **2–4 years initial + extensions**
         Production   **From 20 to 25 years** from production start-up

Relinquishment: At end of each main exploration phase, 25% of original area relinquished. At end of exploration phases, some governments require total relinquishment of all but development areas.

Exploration Obligations: Wide variations in amount and timing

Bonuses: Mostly in highly prospective areas as an *equalizer*

Rentals: Usually amount to very little

Royalty: Most countries have a royalty of some sort. World average is 7%. General range is 0–20%.

Cost Recovery Limit: A phenomenon of PSCs but about **20% of PSCs do not have a limit.** World average is **60–65%.** General range is **50–70%.**

Depreciation: Not required for cost recovery in 40–45% of PSCs. World average depreciation is **5 year SLD**

G&A Expenses: Usually a formula based on Capex and Opex (if allowed)

Profit oil split: PSC phenomena; about **80% are sliding scale**; most are based upon tranches of production

Taxation: Almost all systems have direct or indirect taxes. Equivalent of corporate income taxes averages **35%.** Others usually include withholding taxes at **15%±**

Ringfencing: 60% of the countries in the world are ringfenced; **20%** have a modified ringfence with some relief; **20%** have no ringfence, primarily royalty/tax systems.

Domestic Market Obligation: Usually a formula, fairly rare, that dictates a percentage of entitlement that must be sold to the government at a reduced rate from world price

Government Participation: About 40% of governments have the option to *back-in* at discovery or commerciality; percentage back-in ranges from **10–15% to 50% ±**; for the countries that have the option, the **average is around 30%.**

Dispute Resolution: Mostly international arbitration.

Other: There is nearly always something else to consider. For example, see *crypto taxes* in the next section.

Crypto Taxes

A tax is a compulsory payment pursuant to the authority of a foreign government. Fines, penalties, interest, and customs duties are not taxes. Crypto taxes are those things that just don't seem to fit the ordinary definitions of taxes, royalties, imposts, duties, excises, severances, etc. Furthermore, these things seldom if ever get adequate representation in the "take" statistics. A few examples include:

- Data purchase costs/fees
- Local office requirement
- Surface rentals
- Training fees and scholarship funds
- Customs duties—or customs exemptions that don't hold up
- Cumbersome visa requirements
- Social sphere development costs (written or unwritten)
- DMOs
- Mandatory currency conversions
- Unusually low procurement limitations
- Hiring requirements
- Hostile audits
- Government cost recovery
- Excessive (government-owned) pipeline tariffs
- Price cap formulas
- Short loss C/F periods
- Performance bonds
- Value-added taxes (VAT) or goods and services taxes (GST)
- Reinvestment obligations
- Asset based taxes (ad valorem)
- Inefficient allocation mechanisms—slow indecisive awards
- Unrealistic permitting and impact statements
- Oppressive government controls
- Contract official language—other than English

COMMENTARY

Although the time and costs can be overwhelming, the international sector offers plenty of excitement for companies venturing into the international sector. The most dramatic differences are in the size of the geological structures, the budgets, and, of course, the terminology.

The science of fiscal system analysis and design is relatively new and thus some lack of standardization of terminology and methodology is understandable. But too many times misunderstandings occur because of the diversity of terminology. Hopefully, we will make progress developing standards.

Take statistics that attempt to capture the important issue of the division of profits are useful but have weaknesses. It is important to be aware that the main weaknesses are that bonuses just do not adequately get captured in the statistics.

5

Current Developments in PSCs

Today's climate is characterized by companies and governments awakening to a new era. Countries have become more proactive in their efforts to compete with each other for capital and technology. As an example, in November 1996 for the first time ever, Pertamina (the NOC of Indonesia) went *on-the road* and gave promotional presentations to industry in Houston and London, which focused on their country's geological prospectivity and their contract terms.

CONTRACT TERMS AND PROSPECTIVITY

Contract analysis from a financial point of view must be closely linked to the geological prospectivity associated with a given license area or region. This is not a new concept. Adam Smith in *The Wealth of Nations* (1776) characterized both agricultural acreage, as well as mineral deposits (coal, copper, gold, precious gems, etc.), in terms of fertility and situation. Agricultural fertility equated with, for example, the richness of the soil. For extractive deposits such as coal, fertility was a function of the thickness of the overburden, the quality of the coal, and so forth. Situations dealt with the distance from market and the relative costs of transportation.

Smith pointed out that with coal mines, both fertility and situation were important, but the same was not true for a gold mine. Fertility is important for gold mines (richness of the ore, etc.), but situation is not so important. The transportation costs per unit are relatively low compared to the value of this commodity. The opposite is true with coal.

	Fertility	**Situation**
Coal or gas	Important	Important
Gold or oil	Important	Not so important

These same concepts apply to the petroleum industry. Fertility (prospectivity) for both oil and gas is important, but their situations are not identical. Gas is much more sensitive to distance from market than oil. This is why half of the world's nearly 5000 TCF gas is stranded. It is too far from markets. Of course, half or more of the world's conventional oil reserves are non-producing, but not for the same reason.

A catalogue of contract terms and geological prospectivity is depicted in Table 5–1. Tough terms usually correlate with good rocks. The various elements that capture the essence of *tough terms* and *good rocks* are summarized in this table. There must be a balance, of course, between the fertility and situation and the associated terms. There is much involved in this relationship.

Table 5–1 The Balance Sheet

Prospectivity	Contract Terms
• Expected field size distributions	• Type of system: PSC, service
• Petrophysical characteristics:	agreement, R/T system
porosity, permeability, hydrocarbon	• Signature bonus
saturation, etc.	• Working program: seismic and
• Well deliverability	relinquishment, drilling expenses,
• Estimated success probability: source,	timing, bank guarantees
seal, reservoir, migration, etc.	• Royalty
• Oil vs. gas: fluid properties, API gravity,	• Cost recovery limit
wax, HS_2, etc.	• Effective royalty rate
• Data: quality and quantity	• Government take
• Exploration drilling costs	• Government participation
• Post discovery costs: development	• Entitlement
drilling, production facilities,	• Cost savings index
transportation costs, operating costs	• Ringfencing
• Water depth and climate	• *Crypto* taxes
• Political risk	• Contract stability

RISK AND REWARD

There must also be a balance between risk and reward. The industry standard evaluation tool is the expected value (EV) approach—also known as EMV—which yields a *risk weighted* value as shown in the following equation.

EMV (EMV)

$$EMV = (Reward * SP) - [Risk\ capital * (1 - SP)]$$

Where:
EMV = EMV
Risk capital = Bonuses, dry hole costs, G&G, etc.
SP = Success probability
Reward = Present value of a discovery based on
discounted cash flow analysis discounted
at corporate cost of capital.

This formula is the foundation of risk analysis and decision making. The decision rule is that EMV must be positive to consider making an investment, i.e., the risk-weighted potential reward outweighs the risk.

Typically signature bonuses and work commitments capture the essence of the risk side of the equation. Nearly all of the other elements on both sides of the balance sheet affect the reward side of the equation if a discovery is made. The linkage between risk and reward, then, is the probability that one outcome or another might occur.

Two key elements in the exploration business are estimates of success probability (sometimes called *chance factor*) and the anticipated or target field size. Post mortem analysis of exploration efforts of the past couple of decades indicates that explorers have been optimistic in their estimates of both probability of success as well as field size distribution. The rates of success have not been as robust as expected; and when discoveries have been made, they typically have not been as large as expected.

RESERVE REPLACEMENT

Companies have managed to replace reserves—but only partially through exploration. The demands are great. Wall Street pays close attention to reserve replacement ratios and finding costs. This creates intense pressure on companies to *book* barrels—regardless of the *value* of those barrels. Mobil Corporation provides a good example of how difficult it would be to replace reserves through exploration alone. Mobil (prior to the Exxon acquisition) produces around 1.75 million barrels of oil equivalent per day (BOE). Thus, Mobil would have to find at least 640 MMBOE a year in order to replace production through exploration alone. Just a couple of 300+ MMBOE discoveries per year or so. This is just not happening, and

Mobil is a typical example among many. With the Exxon/Mobil merger, the new organism will be producing more like 4.45 MMBOE per day or 1.6+ Billion BOE per year.

These are obviously dramatic times in the far upstream end of the industry, but exploration is simply not what it was even a generation ago. More than 80% of the world's oil production comes from fields discovered prior to 1973.[1] Giant discoveries are not a thing of the past, yet they are extremely rare these days as shown in Table 5–2. And, just as the industry appears to be facing the reality of a maturing planet, the mega mergers are changing our landscape. In the past few years, as the industry has been coming to terms with the realities of exploration business, another dynamic has evolved.

Table 5–2 Large Field Discoveries Worldwide*
Greater than 50 MMBOE (excludes U.S. and Canada)

Discovery Size	Number of Reported Discoveries			
MMBOE	1960s	1970s	1980s	1990s
50–100	235	261	300	314
100–200	105	162	113	90
200–500	179	208	170	154
500–1,000	90	95	66	52
1,000	129	116	90	20

* From: Peter Rose, "Analysis is a Risky Proposition," AAPG *Explorer*, March 1999. Based on Petroconsultants data 5/96 [1990s data extrapolated].

Exploration acreage is taking on more and more of the characteristics of a commodity. This is because of the dramatic increase in competition among companies for exploration and development opportunities, as well as the competition among countries for exploration capital and technology. There are more companies than ever before seeking opportunities worldwide, and there are more countries than ever before open for business.

The market for projects and acreage is much more competitive and efficient. Governments are acutely aware of what the market can bear, and the terms companies are getting are not nearly as good as they were 20 years ago—not relative to dwindling prospectivity. Furthermore, governments are demanding more aggressive and faster relinquishment of acreage so that they can turn the acreage over more quickly than in the past.

Each year there are 40 to 50 countries offering official license rounds or *blocks offers*. Those countries with official license rounds at year-end 1998 are listed in Table 5–3. Out of this group of countries, nearly a third were not *open* even 10 years ago. In addition to the official license rounds, there are many countries that entertain offers and negotiations *out-of-round*. Each year, approximately 20 countries make major changes to their petroleum fiscal systems, and more countries than that introduce new petroleum laws, model contracts, or regulations.

Table 5–3 Countries with Official Block Offerings at Year-end 1998*

Latin America	Argentina, Bolivia, Brazil, Colombia, Cuba, Falkland Islands, Guatemala, Nicaragua, Trinidad, Tobago
Europe	Bulgaria, Denmark, Faroe Islands, Hungary, Ireland, Netherlands, Norway, UK
FSU	Kazakhstan, Russia, Tatarstan, Yakut-Sakha, Uzbekistan
Africa	Algeria, Angola, Benin, Cameroon, Egypt, Equatorial Guinea, Gabon, Madagascar, Namibia, Nigeria, Senegal, South Africa, Togo
Middle East	Pakistan, Iran, Iraq, Qatar
Far East	Australia, Brunei, Cambodia, China, India, Indonesia, Mongolia, Nepal, New Zealand

* From: *AAPG Explorer*, August 1998 pp. 12–17.

Exploration results in recent years have not been as successful in terms of the number and size of discoveries. As a result, the industry is moving into higher-cost environments. Furthermore, terms are tough; countries now extract resource rent much more efficiently and effectively. Unfortunately some companies interpret this trend as a measure of government greed. In most cases, greed is not the issue; we are seeing an increasingly efficient and competitive marketplace at work.

COMMENTARY

The future of exploration is not dead, but companies must go into deeper water and more remote, inhospitable frontier regions, both geographically and politically. Advances in technology have been spectacular, but this is because of necessity due to lower prices, tougher terms, deeper water, and smaller, more subtle traps. The business of petroleum exploration has always been a high-risk business. But in many respects, it is tougher these days.

REFERENCES

[1] Laherre, J., "Production Decline and Peak Reveal True Reserve Figures," *World Oil*, December 1997, p. 77.

6

The International Gas Industry

International exploration for oil and gas can be exciting and rewarding. Yet, although oil and gas are closely related, as commodities they can be as different as soy beans and gold. Few business ventures can be as thrilling as an oil discovery, and few things are as disappointing as a gas discovery in the wrong place. For all practical purposes, oil and gas are two separate industries. These three key statistics capture the differences:

- **Nearly 10 BCF (10,000,000 MCF) of gas is being flared daily worldwide.** This is nearly enough gas to feedstock all existing LNG trains worldwide. Currently, *no* oil is being flared. In the famous Russian Samotlov oilfield alone, gas is flared at rates of up to 2 BCFD. That would be enough to heat 4 million average three-bedroom homes (in the United States) at approximately 500 cf/d or .5 MCF/day each—or 18 million Russian homes.

- **Roughly 87% of world's 225 BCFD (*O&GJ Databook*, 1998) gas production is intranational (not international).** Of the 13% that crosses an international border, only about a third of that, or 4.4% *(BP Statistical Review of World Energy*, 1996), is liquefied (LNG) and shipped to market.

- **Half of the world's discovered gas (about 5,086 TCF) is *stranded.*** This means it is too far from market to justify the pipe needed to transport it economically.

Typically gas discoveries do not get developed until a long time after discovery. After nearly twenty years, the ARCO-operated Northwest Java (NWJ) block just off the coast from Jakarta, Indonesia, a $1 billion-plus gas development project, finally got underway in the early 1990s. Nonassociated gas (gas not associated with oil production) had simply been shut-in, and the associated gas up to that time had to be flared. The NWJ block boasts the first official PSC ever signed, on August 18, 1966. In NWJ and the adjacent Southeast Sumatra (SES) block, more than 70 BCF of gas has been flared since those contracts were signed nearly 30 years ago. SES was the second PSC signed in Indonesia.

Figure 6–1 illustrates the timing differences between oil and gas operations in terms of start-up and rate of production. The statistic that captures rate of extraction best is the ratio of peak production to ultimate recovery. For example, a typical large oilfield in the North Sea might produce roughly 10% of ultimate recoverable reserves in a peak year of production— this is a production-to-reserves ratio (P/R) of 10% (which is 10% of total recoverable reserves in a peak year of production). For smaller fields in Indonesia the P/R ratio might be more like 20% or so. For many developed gas fields worldwide, the P/R ratio might be more on the order of 3–4% or less.

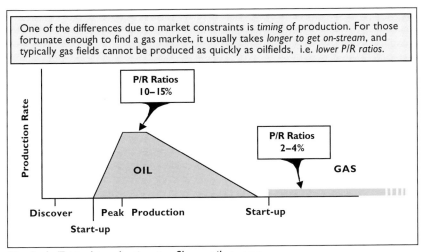

One of the differences due to market constraints is *timing* of production. For those fortunate enough to find a gas market, it usually takes *longer to get on-stream*, and typically gas fields cannot be produced as quickly as oilfields, i.e. *lower P/R ratios.*

Fig. 6–1 Typical production profiles—oil vs. gas

The difference between technical success and commercial success is development threshold field size. If the development threshold for oil is 25 million barrels, then the probability of commercial success is substantially less than that for technical success. Anything smaller would be uneconomical to develop. Imagine the difference between technical success and commercial success if development threshold for a gas discovery is 500 BCF to more than 3 TCF (100–500 MMBOE). Imagine too, if the field development threshold is that large, how much larger the exploration threshold must be. One reason exploration thresholds for gas are usually an order of magnitude greater than for oil is because of the risks. Companies must look for *at least* the 300 to 500 MMBBL fields. If a smaller size field is found—depending on the economics—it is either developed or agonized over for a few years. With gas the agony is usually greater, as is the lead time. Gas can be a real disappointment.

The Gas Curse

There is a saying in many parts of the international oil patch: *What's worse than a dry hole?—A gas discovery!* At least a dry hole does not ordinarily incur the added costs of extended testing. Furthermore, when a company is fortunate enough to develop a gas discovery, it is usually not as valuable as it would have been had it been an oil discovery.

Table 6–1 compares the volume of recoverable gas vs. oil for a hypothetical drilling prospect. All volumetric parameters are shown. The main difference is in the recovery factor. Typical recovery factors are used—gas is simply more mobile, and recovery factors often twice that for oil under primary recovery are not unusual. Nevertheless, if the reservoir holds oil, the recoverable reserves are on the order of 400 MMBBLS. However, if the reservoir holds gas, the recoverable reserves would be on the order of 1300 BCF of gas or about 220 MMBOE (using a thermal conversion factor of 6 MCF per BBL). A detailed summary of the calculations is provided in Tables 6–2, 6–3, and 6–4.

Table 6–1 Volumetric Comparison

	Gas	Oil
Reservoir depth (ft)	9,000	9,000
Drainage area (acres)	5,000	5,000
Zone thickness (ft)	200	200
Porosity average	25%	25%
Hydrocarbon saturation	75%	75%
Pressure gradient (psi/ft)	0.44	
Initial formation pressure (psi)	3,960	
Gas compressibility factor (Z)	0.93	
Oil formation volume factor		1.25
Reservoir temperature(R)	740	
Initial in-place volume	1,662 BCF	1,164 MM
Recovery factor	80%	35%
Recoverable reserves	1,330 BCF	407 MM
Thermal parity (6:1)	222 MMBOE	

Table 6–2 Gas Volumetric Estimate

$$Vi = \frac{43,560 * A * h * Phi * Sh * Pi * Ts}{Ps * Zi * Tf}$$

Where:

43,560 = Square feet per acre or cubic feet per acre-foot

Vi = Initial in-place gas volume (standard cubic feet, standard pressure and temperature)

A = Drainage area (acres)

h = Pay zone thickness (feet) or also called reservoir thickness

Phi = Porosity (decimal)

Sh = Hydrocarbon saturation, decimal (percentage of rock pore space filled with hydrocarbons).

Pi = Initial reservoir pressure (pounds/inch2 [psi])

Ts = 520° Rankine (standard temperature 60°F + 460°)

Ps = 14.7 psi (standard pressure)

Zi = Initial gas compressibility factor (see examples in Table 6–9)

Tf = Reservoir temperature in Rankine (F° + 460°)

Recoverable gas is determined by multiplying *Vi* by an estimated recovery factor or calculating remaining gas volume at abandonment (*Va*) and subtracting *Va* from *Vi*. *Va* requires a separate Z calculation (*Za*) which is a function of abandonment pressure (*Pa*).

Table 6–3 Example Z Values

Pressure psi	Z	Reservoir depth*
500	0.93	1,150
1,000	0.88	2,270
2,000	0.84	4,550
3,000	0.87	6,800
4,000	0.99	9,100
5,000	1.04	11,350
6,000	1.10	13,650
7,000	1.17	15,900
8,000	1.24	18,200
9,000	1.31	20,450
10,000	1.38	22,700

*Assuming normal pressure gradient of around .44 psi/foot

In most reservoirs, pressure (psi) is around 0.44 psi/foot, an ordinary pressure gradient. Thus for example a reservoir with a pressure of 5000 psi would likely be around 11,400 feet deep. The normal range is from 0.433 to 0.465 psi/foot.

Under high-pressure regimes, pressures can approach 0.8 psi per foot and more. Z is also a function of temperature—the example here assumes a reservoir temperature of 300°F and a fairly typical gas with a specific gravity of 0.68 grams/cubic centimeter.

Table 6–4   Oil Volumetric Estimate

$$Vi = \frac{7{,}758 * A * h * Phi * Sh}{FVF}$$

Where:

7,758 = Barrels of volume per acre-foot

Vi = Initial in-place oil volume
(Barrels) (standard pressure and temp)
Also called stock tank oil in place (STOIP)
Also stock tank oil initially in place (STOIIP)

A = Drainage area (acres)

h = Pay zone thickness (feet) or also called reservoir thickness

Phi = Porosity (decimal)

Sh = Hydrocarbon saturation (decimal) [percentage of rock pore space filled with hydrocarbons]

FVF = Formation Volume Factor—This accounts for shrinkage of oil volume as gas comes out of solution at lower surface (standard) pressures and temperatures. Typically FVF is equal to roughly 1.05 plus 0.05 for every 100 cubic feet of gas per barrel, i.e., the gas oil ratio (GOR). For example if GOR is 500 cubic feet per barrel, FVF should be roughly 1.05 + (0.05 * 5) or 1.3.

Recoverable oil is determined by multiplying *Vi* by an estimated recovery factor.

Unless they are quite rich in liquids, close to an existing market, or very large, gas discoveries are often simply noncommercial. When a gas discovery is made it is customarily followed by a well-known ritual that begins with:

"What can we do with all this gas?" [Hand wringing.]

"Imagine how valuable this discovery would be if it were on the United States Gulf Coast!!" [A moment of euphoria.]

"In terms of BOE this is a huge discovery!" [Another short-lived moment of euphoria.]

"Where is the closest market?" [Eyes searching the horizon.]

"What are we going to do with all this gas?" [Hand wringing.]

So what are the options? The list includes:

- Gas sales—produce it and pipe it to market
- LPG—liquids extraction
- Gas cycling
- Gas-fired power generation
- Methanol
- Fertilizer—ammonia/urea
- LNG

These various development options are summarized in Table 6–5.

Table 6–5 Gas Development Options

	LPG Plant	Gas Cycling	Gas-Fired Power Plant	Methanol	Fertilizer Ammonia/ Urea	LNG Plant
Product	LPGs and condensate gas	Condensate gas is reinjected	Electricity	Methanol Hydrogen	Granulated urea	Liquefied methane & ethane
Threshold field size to consider project (BCF)			1,500	1,000	1,000	15,000
Threshold field size to feedstock (BCF)	300–400	250–400	650	500	600	5,000
Minimum feed gas (MMCFD)	60–80	40-75	85	60	80	770 385/train
% produced/year of project life	5%–10% 10–20 yrs	7% 13 yrs	5%± 20 yrs	5% 20 yrs	5% 20 yrs	4% 25 yrs
Project Life	10–20 yrs	13 yrs	20 yrs	20 yrs	20 yrs	25 yrs
Capacity	60 MMCFD 4000 BCPD	30 MMCFD 1000+ BCPD	500 megawatt	2000 tons/day	1750 tons/day	14,000 t/d 5.5 MM t/y
Market requirements	Local & export	Local	Local grid	Export ship	Local truck	Export LNG tanker
Plant location	Local or port city	Local	Local field	Port city	Local	Port city
Required capital cost $MM*	$50–60	$75–100	$350–400	$250–300	$300–400	$2.5–3,000 grassroots
Lead time	3+ years	2-3+ years	1+ years	4 years	5 years	7–10 years**
Construction	2+ years	2+ years	1 years ±	3 years	3 years	3+ years

* Manufacturing facility costs only. ** Long-term minimum take-or-pay contracs required to start.

Gas Sales

The ideal situation for a gas discovery anywhere would be a local market for the gas at a reasonable price. Europe and North America are about the only places where these things are taken for granted. Most exploration acreage is located a long distance from the kind of markets that would make gas sales a simple matter of laying pipe.

As early as 1983, ARCO knew it had at least 3 TCF of recoverable gas in the Yacheng discovery just south of Hainan Island, China. At that time ARCO could guarantee long-term deliverability of at least 400 MMCFD for more than 20 years. The purchase agreement and development plans waited 10 years. And, it will have been much longer than 10 years from the spudding of the discovery well to first deliveries in Kowloon—480 miles and US $1.2 billion away. First deliveries were made in 1996, 13 years after discovery.

Liquids Extraction

Liquids extraction can range from low volume plants that strip out condensates to large scale facilities that liquefy LPGs. LPGs are primarily de-ethanized propanes or butanes with some pentanes thrown in. Condensates are made up of pentanes and some of the heavier hydrocarbons. Worldwide, gas plant sizes typically range from 100 to 500 MMCFD in terms of inlet capacity, and liquid/gas yields range from 20 to 100+ BBLS per MMCF of LPGs and condensate.

The international sector requires rather large-scale projects, and if gas is rich enough, LPG extraction may be a development option. As a rule, the condensate yield alone must be at least 30–40 BBLS per million, and LPGs (propane and butane) may nearly double that. Of the roughly 50-odd gas discoveries reported each year; worldwide average liquid yield on test is on the order of 45 BBLS per MMCF. Gas streams can range from very dry—around 10 BBLS/MMCF—to very rich, more than 150 BBLS/MCF. The general scale is shown in

Figure 6–2. Typically gas associated with oil production is extremely rich in liquids. Non-associated gases are typically dryer.

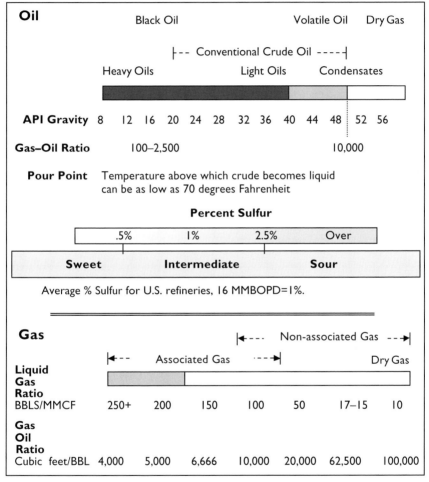

Fig. 6–2 Hydrocarbon spectrum

LPG fractionating plants can be big projects, but they come in a wide range of sizes. At the AMOCO-operated Sajaa gas/condensate field in Sharjah, UAE, up to 450 MCFD was flared at one time at their liquids extraction plant there. Originally there was no market for the residue gas, and it had to be flared.

Gas Cycling

Government policies restricting flaring are becoming more common, but the alternatives are expensive. One alternative is to reinject the gas. Gas liquids projects that reinject residue dry gas are referred to as gas *cycling* projects. Cycling requires additional compressors to recompress the gas and additional wells for injection back into the reservoir. These additional costs are usually prohibitive—liquid yield must be very high to justify these kinds of projects.

Furthermore, with cycling, the reinjected dry gas mixes in the reservoir and, over time, the liquid yield will drop off. The liquid yield curve will ordinarily exhibit a hyperbolic rate of decline. Thus, the economics are additionally burdened by an ever leaner yield for the same amount of inlet gas processed.

Gas-Fired Power Generation

In the mid-1990s, gas-fired power generation boomed. This was because manufacturing facility costs were coming down, efficiencies were improving, and gas enjoyed a reputation as an environmentally correct, clean burning fuel. Furthermore, market demand was robust—although it was dashed somewhat by the economic crisis in the region in the late 1990s known as the *Asian flu*. In 1991, Enron Power Corp. announced plans to build a 380-mw gas-fired power plant at Lawford, England. The project was expected to cost US $300 MM (around $800,000/mw). A similar plant in the late 1990s cost closer to $250 MM.

Earlier in this decade, capital costs ranged from $800,000 to $1,000,000/mw of installed capacity in the United States. Now costs are expected to be more on the order of $650,000/mw. Coal-fired power generation costs about twice as much for plant construction. This is further illustrated in Table 6–6.

Table 6–6 Power Plant Construction Cost Comparison

	Installed Cost (US$B)	Capacity	Time to get onstream
Small gas-fired facilities	$0.7–0.9	10–70 mw	12 months
Large gas-fired facilities	$0.65–0.75	100–500+ mw	18+ months
Coal-fired plants	$1.00–2.00	1,300 mw	6–8 years
Nuclear power plant	$2.00–4.00	300–600 mw	10 years

Simple-cycle efficiency now (single turbine) is around 35–38%. Current efficiencies for combined-cycle power generation depending on turbine size are around 50–55% with projects on the drawing boards for 2005 on the order of 60%. In addition to these efficiencies, electricity transmission and distribution losses are down from around 11% in the 1960s to 5.5% in 1995.

A rule of thumb used in the early 1990s was: *The feedstock requirement MMCFD is equal to the plant capacity in mw divided by 5.* This provided a quick estimate of the feedstock requirement. A 100-mw plant therefore would require around 20 MMCFD feed (100 ÷ 5 = 20). However, conversion efficiencies have increased and the rule of thumb now is: *divided by 6.* Therefore, a 100-mw plant would require around 16.7 MMCFD feed (100 ÷ 6 = 16.7).

METHANOL

Methanol is the alcohol of methane—methyl alcohol CH_3OH. Methane gas is converted into synthesis gas (a mixture of carbon monoxide and hydrogen gas) which is then reassembled into methanol. It is used as a feedstock in the petrochemical industry and can be used as an automotive fuel directly or indirectly. The indirect use is as a feedstock for methyl tertiary butyl ether (MTBE or TAME). It is also used as feedstock for the manufacture of formaldehyde, acetic acid, and other petrochemicals. The capital cost for a world class 2000-ton/day methanol plant requiring up to 60 MMCFD inlet gas is between $250 and $300 million (excluding interest during construction). Unfortunately, methanol prices are as volatile as oil prices.

Fertilizer

Natural gas is also the feedstock for the manufacture of ammonia, the primary feedstock for urea. In early 1994, plans were announced for a world-class ammonia urea complex at Gresik, East Java. Costs were estimated at $242 million for the 1350-metric ton/day ammonia–1400-t/d urea plant. (*Petromin Magazine,* March 1994). This particular plant would not be considered to be a balanced plant. Only about 800 tons/day of ammonia would be required to manufacture 1400 t/d of urea. Ammonia, however, has uses other than as a feedstock for fertilizer. The plant example outlined in Table 6–5 is a balanced plant requiring 80 MMCFD feed gas. The plant is assumed to produce 1000 t/d of ammonia and 1750 t/d of urea.

The Kafco Fertilizer plant in Karnaphuli, Bangladesh came onstream in 1995 and shipped its first cargo of 12,000 tons of liquid ammonia to the United States. The capacity of the $510 million plant is 1500 tons/day of ammonia and 1725 tons/day of granulated urea.

Liquefied Natural Gas (LNG)

LNG is mostly liquefied methane and ethane. Table 6–7 shows an example composition from the Bontong LNG facilities in Badak, East Kalimantan. The problem that faces LNG development is primarily the huge up-front capital cost required to build a full-range LNG chain from gas field development to liquefaction, transportation, receiving terminals, storage, and regasification facilities. The liquefying temperature for LNG is -162C. As a result, processing, storage, and transportation are quite expensive.

Table 6–7 Example Gas Composition Badak LNG

Component	Feed Gas Volume %	LNG Volume %
N_2	.07	.03
CO_2	5.97	-
H_2S	.00	.00
C_1	83.51	90.79
C_2	5.25	5.71
C_3	3.12	2.51
$I-C_4$	.59	.48
$N-C_4$	.69	.48
$I-C_5$	.23	-
$N-C_5$	.15	-
C_6+	.42	-
Total	100.00%	100.00%

A new LNG plant alone can easily cost $2 billion. Existing facilities have a huge cost advantage over grassroots construction. Each expansion train of an existing facility may cost $400–600 million and take from 24 to 36 months to complete. The cost for expansion is nearly half or less that of a start-up. In March 1994, Badak Train-F a 2.3 MMT/year capacity expansion came onstream at Bontong, East Kalimantan. The cost was $522 million, including infrastructure support, housing, and roads. Interest during construction added another $177 million, and the total cost came to $699 million.

A typical cargo for an LNG tanker with 125,000 cubic meters storage is 57,000 tons, equivalent to about 2.7 BCF of liquefied gas. One round trip from Mobil's Arun LNG complex in Indonesia to

Japan takes about 15 days. Assuming 20 cargoes annually, one tanker can transport about 1.14 million tons per year—150 MMCFD equivalent. By the way, it would take at least a 16-inch pipeline to transport this much gas—or to replace the one tanker. Key statistics relating to LNG are summarized in Table 6–8.

Table 6–8 Liquefied Natural Gas (LNG) Facts

• Liquefying temperature for methane: -162°C
• One MCF gas = 43.57 pounds LNG
• One metric ton of LNG = 46 MCF gas
• 140 MMCFD = 1 MM metric tons LNG/year
• 8–9% of feedstock delivered used as fuel to liquefy the rest of the gas

The two-train LNG facility in Table 6–5 producing 5.5 million tons per year would need five tankers at say $250 million each. Regasification facilities and storage at the other end of the line can also add upwards of $700 million.

Comparison of Development Options

Table 6–5 summarizes vital statistics for world-class gas development options. The objective is to provide an indication of the thresholds and boundary conditions that govern or influence international gas projects. Every situation is different, and non site-specific cost estimates are notoriously inaccurate. The difference in construction costs can vary dramatically from one location to another depending on many factors. This is illustrated to some extent by the difference between grassroots construction vs. expansion with LNG facilities previously mentioned.

There is another important difference. Most upstream project costs quoted do not include interest during construction. Most construction cost quotes for gas development and downstream projects include interest during construction. There is a similar difference with operating costs. Upstream operating cost quotes do not include DD&A. When those in the downstream end of the industry quote operating costs, they ordinarily include DD&A and may also include feedstock costs. Regardless of these heavy qualifiers, the attempt here is to give ballpark cost figures for a frame of reference. Interest during construction is not included in the cost estimates summarized in the Table 6–5.

There are additional costs that are not included. Many options require an export terminal, and export products require tankers. The cost for a terminal can range from as low as $200 million to more than $500 million for deepwater port facilities capable of handling large tankers. Also, extended distances from the field to the plant or port facilities will increase pipeline costs.

Table 6–5 shows two field size threshold requirements. One is simply the minimum amount of gas required to feedstock a world class facility for 20–25 years. Yet a discovery would often have to be much larger to go forward with development in order to ensure sufficient feedstock and room for expansion. LNG buyers in the Far East would not even enter into discussions unless there were 12–20 TCF of recoverable gas involved—a different kind of threshold.

Capital cost estimates in Table 6–5 are based upon design, construction, and commissioning of manufacturing facilities only in an environment that is not as benign as the Texas Gulf Coast but not as harsh as the Eastern Indonesian Archipelago where distances are large and infrastructure is scarce. Costs of field development and transportation from field to manufacturing site have not been included.

COMMENTARY

There are numerous countries that are tightening flaring policy, and the option of simply flaring associated gas is becoming rare, but shut-in gas is nearly worthless and so is shut-in oil. The alternative, of course, is to restrict flaring altogether, but then associated oil or liquids would either not be produced or may not be produced economically.

Because of the physical differences between oil and gas that have such a strong influence on profitability, many countries provide better fiscal terms (royalties, taxes, etc.) for gas than for oil. Some countries will clearly define terms for oil while gas terms may be quite vague. Sometimes a contract will contain a simple *gas clause* indicating that if gas is discovered, both parties (government and contractor) will then sit down and discuss/negotiate. The reason is that it can be nearly impossible to anticipate in advance how large or rich in liquids a gas discovery might be, and the options are diverse compared with the relatively simple development options for an oil discovery. Furthermore, the market for gas or gas products is much more complex than for oil. Oil is simply easier to deal with.

The development options in Table 6–5 further illustrate the plight of a gas discovery. First of all, methanol and LNG are simply export products. Thus, the manufacturing facilities must be located near the coast with port facilities. Therefore, additional costs may be incurred transporting the gas from the field to the facilities. And what if a company discovered 5 TCF in a remote part of the world? A world-class power plant, methanol facility, or fertilizer plant would only need about one-tenth of the gas for feedstock more than 20–25 years. Yet 5 TCF would not be enough to comfortably approach a possible LNG buyer.

Fortunately, gas is becoming less of a curse in many places for many reasons. There are some interesting developments unfolding, but change is simply not going to happen quickly.

The Future

Beyond the conventional gas field development options outlined in this book, there are a few potential options on the horizon that could change the face of the industry. The dynamics of some of these will be driven by oil prices. However, the gas industry could be dramatically altered. The most important—or perhaps intriguing—are:

- Fuel cells
- Gas to liquids (GTL)
- Hydrates

It might be worthy to point out that it appears fuel cells and gas-to-liquids (GTL) may soon burst on the scene; just as gas-fired power generation did in the mid 1990s. Furthermore, this may be for the same reasons, namely: improving efficiencies, falling costs, and their environmentally friendly nature. Terminology and classification of natural gas products are summarized in Table 6–9.

Table 6–9 Natural Gas Products

Light Hydrocarbon Series					
C_1	C_2	C_3	C_4	C_5+	Terminology
← LNG →					Liquefied Natural Gas
← CNG →					Compressed Natural Gas
		← LPG →			Liquefied Petroleum Gas
	← NGL →				Natural Gas Liquids
				COND	Condensate
Methane	Ethane	Propane	Butane	Pentanes+	

Source: *International Petroleum Fiscal Systems and Production Sharing Contracts,* Daniel Johnston, 1994

7

Key Concerns of Governments and Oil Companies— Alignment of Interests

It is probably fair to say that the science of fiscal system design and analysis is perhaps not as highly evolved as, say, reservoir engineering, but progress is being made. The most compelling evidence for a bold statement like this is the diversity of terminology involving the most important aspect of fiscal system design—the division of profits. The most common term used regarding the division of profits is *government take*, although there are numerous competitive terms that add to the confusion: *tax take*, *fiscal take*, *state take*, *marginal take*, *financial split*, *bottom-line income split*, *fiscal net*, *rent*, and so forth. This problem was addressed in past issues of the *Petroleum Accounting and Financial Management Journal*, and in Chapters 2 and 4 of this book.

More and more attention (and rhetoric these days) is being placed on the subject of *alignment of interests*. The more efficient and flexible a contract is, theoretically, the more stable. With more and more development (or rather non-exploration) projects, the sharing of risks becomes more aligned as well. However, when it comes to the design of a particular system or arrangement, all countries are different. In the global competition for capital and technology, all countries have their own peculiar boundary conditions, concerns, and objectives that influence the development of policy, strategy, and tactics (*see* Fig. 7–1).

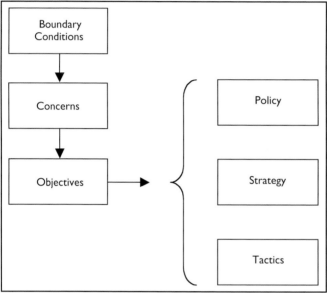

Fig. 7–1 Every country is different

Boundary Conditions

When it comes to severe boundary conditions, few can rival bad geology. However there is an infinite variety of conditions and combinations of circumstances such as being land-locked, perceptions of being gas prone, guerillas in the hills, deepwater, OPEC membership, high-cost over-pressured reservoirs, heavy oil, and so forth. The more severe the conditions, the more flexible and lenient the terms—usually.

Concerns

The political, religious, and cultural foundation of a country can have a substantial influence on the way business is conducted. For many countries, the dire need for capital and technology has overcome their normal aversion to Westerners. The vast potential in Kuwait predicts a major and vicious land rush as they open their doors. However, they have made it clear that there will be no more than four *partners*—perhaps an allusion to the fact that in Kuwait a man may have four brides, which was pointed out by Sheik Sultan, Chairman of Kuwait Oil Corporation, at the Middle East Petroleum and Gas Conference in Bahrain, March 1999.

Another concern of course is contract terms, particularly financial terms. Most governments only want a fair contract that is consistent with what the market can bear—nothing more. The thought of leaving money on the table horrifies them. However, this concern is usually overrated. This is a more legitimate concern for international oil companies (IOCs). They are the ones placed into the competitive role of bidding or negotiating against each other. When money is left on the table, it usually goes to the governments.

Objectives

A country will usually measure objectives in terms of the number of contracts signed and the quality of the work programs associated with those contracts. Work programs are almost always measured in terms of wells drilled. Seismic data acquisition, processing, and interpretation are a means to that end. Governments need wells drilled. If the country is not satisfied with the level of exploration and development activity, then changes must be made to meet their objectives. A lot of time can be lost misjudging the marketplace.

Policy

For many nations, the birthplace of the policies governing petroleum exploration and development is the nation's constitution. Beyond that, some policy matters are dealt with in the country's petroleum laws and regulations. Not all policy, of course, is formally enshrined in the constitution and laws. Some is informal but firmly established. As a matter of policy, some countries do not allow negotiation of contract financial terms and will design the commercial terms themselves consistent with their perception of what the market can bear. Licenses may then be awarded on the basis of a work program bid. Some countries will allow the market to determine what a license area is worth through competitive bidding. Sometimes when a company *wins* a bid, it only means that they have won the right to negotiate further. This may sound a bit unfair, but it happens.

Strategy and Tactics

At this level the NOC personnel usually do their part to try to meet the nation's objectives for the petroleum sector. More and more countries these days are adopting more proactive strategies for marketing their acreage. This is a fairly new trend. However, the typical geo-technocrat has little training in the realm of state-of-the-art international marketing.

For many years, most countries were content to simply wait for business to come to them. Countries are beginning to develop sophisticated (for the oil industry) strategies and slick (for the oil industry) tactics. The age-old *dog-and-pony* shows still abound, and the speech-giving circuit is still active, but NOC personnel realize that there is more to marketing than this. NOC personnel these days are more visible and available, and they are developing marketing plans, *spin* tactics, etc. Considering the amount of money involved, the only surprise is it has taken this long for the far upstream end of the petroleum industry to try to catch up with other industries in this regard. Acreage is not a consumer-based commodity, but it is a commodity, and competition is heating up.

ALIGNMENT OF INTERESTS

A critical aspect of an exploration and production (E&P) agreement is the alignment of the various parties' interests. In most international negotiations, there is considerable lack of alignment prior to contract signing. Obviously, both the IOC and the government want to get as large a share of profits as possible (within reason). However, if the contract is efficiently and appropriately crafted, there should be substantial alignment or *mutuality* of interests as soon as the contract is signed.

There are four key areas of concern:
- Division of profits
- Government guaranteed share of revenue
- Keeping costs down—incentives, cost control, and cheating
- Maximum efficient rate (MER)

Beyond discussion of these important aspects of a contract or fiscal system are the means by which various aspects of *alignment* can be measured and compared. A well-balanced, efficient, and flexible contract will theoretically provide greater contract stability.

Example Fiscal System

The following pages contain example calculations based upon hypothetical contract terms summarized as follows: 10% royalty, 70% cost recovery limit, 60/40% profit oil split (in favor of the government), and a 50% income tax rate.

Division of profits

A key aspect of any contract is how to divide profits. This is determined prior to contract signing. In fact, it is usually the first thing that is agreed upon. If agreement can be reached on the commercial terms, then other terms can be ironed out and finalized. Anything else would be a waste of time. While much of the discussion of the division of profits focuses on *economic* profits (gross revenues less costs associated with obtaining those revenues), timing is everything. A major objective of both parties to these contracts is maximizing wealth—financial profits, not economic profits.

If a discovery is made, an IOC will choose a development plan and operating practices that maximize present value for the company. With extremely few exceptions, this plan will also maximize the value of the government's share of profits, achieving an almost perfect alignment of interests. This should not be surprising: anything less would be extremely inefficient. Government profit share (take) can range from as high as 90% or more (Venezuela) to as low as 25% (Ireland), yet the incentive to maximize present value exists throughout this range.

Part of a well-balanced contract involves matching the division of profits to the prospectivity of a given block or area. Some areas are sufficiently exciting geologically to justify the higher government takes. However, there will always be some field size threshold below which development is not economically feasible—development threshold field size. The difference between commercial success and technical success in exploration is development threshold field size. Technical success is where a discovery is made. Commercial success is a discovery greater than the development threshold which is always greater than zero. As the size of discovery approaches the development threshold, alignment of interests begins to diverge. The greater the sunk cost position at the time of discovery—and the tougher the terms—the greater the divergence.

GOVERNMENT TAKE CALCULATION

Government share of economic profits (full cycle)

A	**100%**	Gross revenues
B	**10%**	Royalty (10%)
C	**90%**	Net revenue
D	**35%**	Assumed costs
E	**55%**	Profit oil
F	**33%**	Government profit oil (60%)
G	**22%**	Company profit oil (40%)
H	**11%**	Income tax (50%)
I	**11%**	Company cash flow
	83.1%	Government take [(B+F+H)/(A-D)]
	16.9%	Company take [I/(A-D)]
	1.31%	R factor (full cycle) [(D+I)/D]
	57%	Entitlement [D+G]

Government Guaranteed Share of Revenues (ERR)

Any level of government take will influence the point at which a field becomes non-commercial; and, of course, the higher the government take, the larger this threshold becomes. However, it is difficult to structure a system in such a way that development threshold field size approaches zero. The main reason is that most governments simply must have a guaranteed share of revenues in any given accounting period, either through royalties or cost recovery limits. The world average guaranteed share of revenues in any accounting period for a government is around 20%. For R/T systems it is less, around 10% or so; and for PSCs, it is closer to 30%. Some guaranteed share of revenues for the government is actually in the interest of both parties.

A government could receive nothing in a given accounting period if the contract or system has no royalty or cost recovery limit. This can happen even with profitable fields during the early years of production when substantial exploration and development costs are being recovered. This could be politically dangerous for a NOC. And if it is dangerous for the NOC, it could be dangerous for the IOC— another form of alignment.

The ERR is defined as the minimum share of revenues the government might expect in any given accounting period (usually ignoring the effects of government participation). In the example fiscal system, the government is guaranteed a minimum of 22% even though the royalty is only 10%. This is because of the combination of the cost recovery limit and the profit oil split. With sufficient deductions, the company would pay no tax in that accounting period.

ERR CALCULATION

Government share of revenues at saturation (single accounting period)

A	100%	Gross revenues
B	10%	Royalty (10%)
C	90%	Net revenue
D	70%	Cost recovery limit (70%)
E	20%	Profit oil
F	12%	Government profit oil (60%)
G	8%	Company profit oil (40%)
H	0%	Income tax (50%)
	22%	ERR [B+F]
		Government Share of Revenues
		(Assuming Unlimited Deductions)

Keeping Costs Down

Oil companies are relatively obsessed with keeping costs down, and in the spirit of nearly perfect alignment on this issue, governments too have huge concerns about keeping costs down. This is because almost all systems, R/T systems as well as PSCs, allow the companies to directly recoup expenditures if sufficient revenues are generated. Under an R/T system, the mechanism is called deductions. The same mechanism exists under a PSC but is called *cost recovery*. Thus governments have a keen interest in seeing costs kept as low as possible, but so do IOCs. In this area, there is a clear alignment of interests, although there are varying degrees of incentive. And it can be measured. One must simply ask:

If costs are reduced by one dollar, who benefits and by how much?

Typically, if a dollar is saved, the IOC will end up with about 33 cents on the dollar. It may not sound like much, but it works. In Indonesia under the standard oil contract, the contractor receives only about 15 cents on a dollar saved. This is one end of the spectrum; the other is upwards of 69–75 cents on the dollar (U.K. and Ireland respectively).

The term *gold plating* often arises in the context of those countries like Indonesia where the savings index is quite low. True gold plating is where a company is encouraged to spend more than it otherwise would because of inefficiencies in the fiscal system. The more they spend, the more they make. However, this kind of arrangement is extremely rare. Systems are not that inefficient.

However, if a company receives only 15 cents on a dollar saved, then the incentive to save is certainly mitigated. Why not have a Rolls Royce for a company car if it only really costs about 15 cents on the dollar? This is an area where the alignment of interests begins to fail. The division of profits in the mid-1990s vintage Malaysian contracts could be very similar to the Indonesian 85/15% split (government take = 85%). However, the savings incentive in Malaysia was closer to 25 cents on the dollar relative to 15 cents on the dollar in Indonesia. The reason is that Malaysia has a 10% royalty and the entire Indonesian 85% take was profits based. When calculating the savings *index* (if you will) only the profits-based mechanisms have an impact.

EFFECT OF SAVING A DOLLAR

Resulting division of revenues
(single accounting period or full cycle)

D	1.00	Assumed cost savings
E	1.00	Profit oil (Increased by $1 worth of oil)
F	.60	Government share of profit oil (60%)
G	.40	Company share of profit oil (40%)
H	.20	Income tax (50%)
	.20	Company share
	.20	Company savings incentive (index) [20 cents on the dollar]

Marginal Government Take

There is another interesting index that provides some insight into fiscal system/contract design and the interests of the parties. It is most commonly referred to as marginal government take although it is not widely used. It can be viewed a couple of different ways, but I prefer the following. Marginal government take is the division of profits resulting from an increase in oil prices. One must simply ask, "All other things being equal, what would happen if oil prices increased by $1?" The calculation is similar to the ordinary calculations of take except costs are assumed to be zero. (All other things being equal, no additional costs would be associated with the revenues resulting from an increase in oil prices.)

Marginal Government Take Calculation

Government share of increase in oil prices (full cycle or single accounting period)

A	100%	Gross revenues
B	10%	Royalty
C	90%	Net revenue
D	0%	Assumed costs (by definition)
E	90%	Profit oil
F	54%	Government profit oil (60%)
G	36%	Company profit oil (40%)
H	18%	Income tax (50%)
I	18%	Company cash flow
	82%	Marginal government take [(B+F+H)/(A-D)]
	18%	Company take [I/(A-D)]

Remember government take was 83.1%. Marginal government take is always less if there is a royalty.

Cost Control and Cheating

There appears to be a morbid fear that oil companies will waste the natural resources through frivolous spending as well as cheating on costs, i.e., claiming they spent more than they spent. In my opinion, governments hold most (but not all) of the cards on this issue. Their primary means of protection include:

- The budget process—AFE (AFEs)
- Procurement regulations and thresholds
- Audits
- Laws and penalties for non-compliance or fraud

- Natural inclination of companies to economize due to profit sharing mechanisms in all systems (savings index)
- Partners–watchdogs

There could be substantial gains for a company from cheating on costs. For example, if a company could recover costs that were actually not spent, then this could provide a windfall resulting in a clear lack of alignment of interests. However, this is not so easily done. Companies are like people. Some (perhaps around 15%) will never cheat. Many, however, require some oversight to ensure that there is little or no major cheating. Even so, some (perhaps around 10%) will always try to cheat.

MER

The conventional wisdom in almost all governments (aside from the odd petroleum engineer in the NOC) is that an IOC's only concern is on short-term profits and that it will produce oil too quickly and inefficiently and therefore leave behind more than they otherwise would.

The concept of the MER is an emotional issue with many. There is a belief that a strong relationship exists between rate of extraction and ultimate recovery. Unfortunately, terms such as *rape and pillage* are used in regard to oilfields that are supposedly produced too quickly. For gas fields, the terminology is expanded to *rape, pillage, and burn* in reference to the roughly 10 BCF of gas that is flared daily worldwide. If true, this would constitute a huge misalignment of interests. However, most scientists believe that there is not a dramatic relationship between rate of extraction and ultimate recovery.

An index that captures rate-of-extraction is the production to reserves ratio (P/R). This index provides an excellent and useful measure. It represents the percentage of total ultimate recoverable reserves produced in a peak year of production for a given field. For example, the North Sea Thistle field produced 45 MMBBLS at its peak in 1982. Ultimate production is projected at around 450 MMBBLS. Thus the P/R ratio for Thistle is 10%. For large oilfields (outside of the Middle East), this is fairly typical. For smaller oilfields on the outer continental shelf of the U.S. Gulf of Mexico, a P/R ratio of 15% is more typical. Many Indonesian oilfields have P/R ratios of 20–25%. The production decline rate following the peak years of production is often close to or greater than the P/R ratio. For example, if a field produces 25% of its reserves in one year, the decline rate in following years will typically be greater than 25%. Thus even high rate Indonesian fields will produce for more than 20 years. The question arises though: *Are the Indonesians allowing the IOCs to produce too quickly?*

Research indicates that this is not highly likely—in fact on this issue, there is a strong alignment of interests. This research consists of a detailed summary of approximately 30 papers/books that deal with any aspect of the relationship between rate of extraction or well spacing and ultimate recovery.

Are reservoirs *rate sensitive?* Certainly all reservoirs and fields are different, but the general consensus of the research from a purely technical (non-financial) point of view is summarized below.

Water Drive Reservoirs—Rate Sensitive

The faster the reserves are depleted, the higher the recovery factor but not by much. This is particularly true of gas production where faster production results in less bypassed gas. As far as gas is concerned, there appears to be nearly universal agreement. To a large extent, the same is true of oil production, but there is no universal agreement regarding oil as there is for gas as far as water drive production. However, the general consensus is that even with oil, the recovery factor will likely be greater the faster the oil is produced. The recovery factor may not be dramatically higher, but it will be higher because there will be less oil bypassed by the encroaching water.

Solution Gas Drive Reservoirs—Not Rate Sensitive

The reservoir will produce the same reserves regardless of the number of wells drilled or rate of extraction. There appears to be fairly wide (but not universal) agreement on this particular conclusion.

Gravity Drainage Reservoirs—Rate Sensitive

The slower the reserves are depleted, the higher the recovery factor. However, very little of the world's production comes from gravity drainage.

Beyond the purely technical perspective, as soon as present value discounting is factored-in, the issue of rate versus recovery becomes a small issue or perhaps a non-issue. Faster is better up to a point. It usually depends upon the number of wells a company is willing to drill, and this is determined primarily by which scenario maximizes present value. Certainly there are circumstances where waste may occur or poor reservoir management can be found. However, these are rarely *alignment* issues; they would more likely involve a lack of technical competence or negligence.

The mandate of both governments and IOCs is to maximize wealth not recovery. This is not a harsh view.

This issue drives to the heart of the wealth of nations and corporations. Oil-in-the-ground is like money in the bank but in a non-interest bearing account. If a country can monetize their mineral wealth, then this capital can be converted to other uses that enhance the nation's wealth—roads, schools, hospitals, technology, etc.

REFERENCES

Johnston, D., *International Petroleum Fiscal Systems and Production-Sharing Contracts,* Course Workbook, 1999.

Johnston, D., *Economics and Politics of Energy,* Course Workbook, 1999.

Johnston, D., *International Petroleum Fiscal Systems and Production-Sharing Contracts,* Tulsa, Oklahoma: PennWell Books, 1994, p. 65.

Johnston, D., and Johnston, D., *Maximum Efficient Production Rate,* University of Dundee, Scotland, 2002.

8

Fiscal System Design— the Ideal System

If asked by a newly created government to provide advice on grassroots development of a petroleum fiscal regime, what would be the ideal system? In many countries, preexisting laws, regulations, taxes, and contracts make this an academic exercise. In most countries, only certain elements can be modified, or changed, but given a clean slate, what would be an ideal regime?

There appears to be fairly clear agreement among academics and practitioners alike on the criteria for an effective, efficient petroleum fiscal system. The ideal regime should:

- Ensure a stable business environment and minimize sovereign risk

- Discourage undue speculation

- Provide potential for a fair return to both the state and to companies, balancing risk and reward

- Avoid complexity and limit administrative burden (on both the state and companies)

- Allow enough flexibility to accommodate changes in perceived prospectivity, and economic conditions

- Promote healthy competition and market efficiency

The proposed system design is based on the assumption that the government has sufficient faith in the NOC or Ministry to grant it sufficient authority to institute such a system, negotiate terms, and award licenses.

For example, Ecopetrol the NOC of Colombia for many years knew the inflexible statutory 20% royalty for oil had become obsolete (it was too high under almost all circumstances). However, the royalty rate was imbedded in petroleum law. Ecopetrol simply could not change it or negotiate lower rates as it deemed fit as many other NOCs are able to do. Only recently has the royalty rate been changed in Colombia and it is more competitive, but time was lost.

Key elements of the proposed design are summarized as follows:

Summary of proposed terms

Type of System:	PSC
Allocation Mechanisms:	Sealed bids on specific blocks and direct negotiations
Work programs:	Biddable or negotiable
Duration and Relinquishment:	Biddable or negotiable
Signature bonus:	Nil
Production bonus:	Production start-up bonus $1 MM
Royalty:	Nil
Cost recovery limit:	50% (Unrecovered costs are carried forward)
Profit oil split:	Biddable with a sliding scale linked to R Factor or ROR

(Two examples)

Government Share

R Factor*	ROR*	(bid items)
0–0.5	0–10%	X_1
0.5–1.0	10–20	X_2
1.0–1.5	20–25	X_3
1.5–2.0	25–30	X_4
2.0–2.5	30–35	X_5
> 2.5	> 35%	X_6

* either one or the other

Profit gas split: Biddable, same as for oil

Taxation: To be paid out of the government's share of profit oil/gas

Government participation: 10% carried through confirmation of discovery

Customs duties: Exempt

Dispute Resolution: Binding international arbitration

Ringfencing: Yes—no consolidation allowed

Production-Sharing Systems

From a financial point of view, there is little magic in using a PSC, because PSCs have philosophical appeal to most governments and citizens and a political correctness that is taken very seriously these days in many countries. The PSC in the eyes of many is the perfect alternative to the hated old *concessions*. The most distinguishing features of a PSC relative to concessionary (R/T) systems deal with extended government ownership and greater sovereign control. The mechanical and financial differences can be very small. However, an oil company's share of profits (take) under a typical PSC is about half that under a typical concessionary system. Thus, for various reasons, the PSC has become the system of choice these days for most countries in the process of opening up new acreage or remodeling their systems.

Allocation Mechanism

From the government point of view, competitive bidding is most advantageous. Companies would prefer direct negotiations. The proposed system would award licenses primarily through sealed bids on the basis of a profit oil split bid, but it would also allow direct negotiations for certain blocks at the discretion of the NOC. The combination of sealed bidding on specific blocks and/or direct negotiations can provide the NOC substantial flexibility. In addition, the work program would also be biddable or negotiable. By providing a profits-based bid item, the burden of *fiscal marksmanship* is removed from the government. Fiscal marksmanship requires a clear understanding or knowledge of what the market can bear under a variety of conditions. By allowing companies to bid on blocks, the government simply grants the rights to the bidder who places the highest value on those rights. The oil companies determine what the market can bear.

In the mid 1990s, the Government of Trinidad created a special group to evaluate appropriate terms for upcoming deepwater block offerings. However, the group was disbanded because oil companies were coming to Trinidad making offers that exceeded the Minister's (and the group's) expectations. The government basically decided to let the marketplace do its work.

Work Program

In many fiscal regimes, winning bids are determined solely on the basis of the work program. This is particularly true of the U.K. and Australia, for example. In these situations, part of the work program bid can take on many of the characteristics of a signature bonus. This happens because companies must commit to some *extra* work (above and beyond what might be considered *technically* appropriate) in order to compete. At any given time, any license or block has some amount of work in terms of seismic data acquisition and/or drilling that would be appropriate. While technical personnel would certainly debate what this appropriate amount of work might be, they generally agree that a competitive work program bid must exceed it.

Determination of the appropriate work program should be the domain of oil companies. Yet the bulk of the competition should focus on a profits-based mechanism, not additional work.

Duration and Relinquishment

Companies need sufficient time for exploration, market development (in the case of gas), and production operations. Who is more qualified than the oil companies to determine how much time would be appropriate? This should be a bid item.

A typical schedule would provide 6–8 years for exploration in 3 exploration periods with 25% relinquishment of the original area in a contiguous block of acreage after each of the first 2 phases of exploration. After that, all but development areas associated with discoveries should be relinquished. Duration for production should be a minimum of 25 years for oil. Typically too, many countries will provide an additional 5 to 10 years market development phase for a gas discovery. Ten years should probably be a minimum for regions with no gas market or infrastructure.

Bonuses

Signature bonuses are used in about 40% of fiscal systems, but they have a strong negative impact on exploration economics and are particularly discouraging to smaller companies. Companies prefer to spend their limited exploration funds on data acquisition, and this is also consistent with most government objectives.

Production or start-up bonuses, payable on commencement of production, are more acceptable to companies and may be useful to help government gear up to meet its increased regulatory workload. Production bonuses are so much more benign than signature bonuses because they reside on the *reward side* of the exploration risk/reward equation. They are only paid if a discovery is actually made and then developed.

Royalty

Royalties are extremely common, but they are also regressive and unpopular with oil companies. The one thing that royalties do most effectively is to guarantee the government a share of production each and every accounting period. However, this can be done more efficiently with a cost recovery limit.

Cost Recovery Limit

A cost recovery limit in conjunction with a profit oil split will ensure that in each accounting period, the government will get a share of production. A royalty could do this but cost recovery limits are less regressive than royalties. A 50% limit is slightly less than the 63% world average, but would not likely be considered to be onerous. Besides, some of the effect in terms of the rate of cost recovery and the ERR would be determined by the oil companies through the biddable/negotiable profit oil (or gas) split.

Profit Oil Split and Tax

In my opinion, it would be best to have companies compete/bid primarily on the basis of a *profits-based* mechanism tied to either an *R* factor (payout formula) or internal rate of return (IRR) thresholds

(ROR systems). Thus the *equalizer* would be a progressive mechanism placing determination of ultimate government take in the hands of the oil companies. There are many advantages to the government (or NOC) with this approach.

Production-based sliding scales typically will provide the government a greater share of profit oil at higher rates of production. Unfortunately—unlike R factors and ROR systems—these scales are totally unresponsive to fluctuations in oil prices. Thus, from a theoretical point of view, R factors and ROR systems are more efficient and more flexible. Yet from a practical point of view, governments strongly prefer production-based sliding scales—about 75% of the PSCs worldwide have them.

By making the profit oil split a bid item, money left-on-the-table and/or winners curse, a phenomenon of competitive bidding would accrue to the benefit of the government. If no licenses are awarded, it would be difficult for citizens, the press, parliament, or oil companies to claim that the NOC had created unrealistic or unbalanced terms—this is left up to the companies. Let the oil companies and competition determine what the market can bear. This is what oil companies do when they procure most of their goods and services from the petroleum service industry. It is very efficient.

It is proposed that taxes be paid for and on behalf of the companies out of the government share of production. This kind of arrangement is found, for example, in Oman, Egypt, Syria, the Philippines, and Trinidad. Contracts with taxes treated this way are some of the most stable arrangements in the world. All fiscal elements then are embodied in the contract, and the contractor would not be affected by government changes in tax rates. If structured properly, these taxes *in lieu* can be treated in the same way as if they had been paid directly by the oil company for home country tax credit purposes.

Government Participation

Government participation as a working interest member of the contractor group is not popular with petroleum companies for a variety of reasons, including reduced entitlement, undue government influence in technical and operating committee meetings, and reduced company take. However, the inclusion of a small direct interest (say 10%) would not dramatically impact project economics, and would help train key government staff and give government insight into how the industry does business.

In approximately 50% of countries, the NOC has an option to *back-in* and take up a working interest in a discovery. The average NOC working interest in these countries is 30%. Many NOCs have the option of funding their working interest share of cash calls from their share of production, up to a certain percentage. This helps avoid the potential problem of the NOC not being able to meet cash calls.

COMMENTARY

There is a dramatic variety of systems worldwide. There are some less-than-ideal, poorly designed systems. The systems that are more likely to fall into this category are those:

- With rigid inflexible "fixed" terms (with no negotiating or bidding on elements that effect take)
- With no royalty or cost recovery limit (i.e. no ERR)
- Where licenses are awarded solely on the basis of work program bids
- With strongly regressive features

The U.K. has each of the first three criteria. In the U.K., licenses are awarded on the basis of work program bids, and the commercial terms were almost entirely captured in a single profits-based 30% income tax (prior to 2002). This, by the way, was the same tax that any other industry in the country paid. Even bakeries, shoe stores, and restaurants paid this tax. Just across the international boundary, the effective tax rate in Norway is closer to 75%. The U.K. has an R/T system but the royalty rate is zero. With no royalty and no cost recovery limit, the ERR is zero. (In 2002 the U.K. added another layer of tax of 10% to the petroleum industry.)

Regressiveness is a more common problem. When oil prices increased so robustly in 1999, government take in most countries (approximately 75%) simply went down. This is because most petroleum fiscal systems are regressive, particularly when it comes to fluctuations in oil price. The biggest culprit is royalties. Just about the only systems where government take actually increases with increased oil prices are those with either an R factor, an ROR feature, or a depletion allowance.

Fiscal design must be country specific, and there are often many trade-offs. Yet the design outlined here should accommodate a variety of conditions including shallow vs. deepwater, high vs. low prospectivity, different cost environments as well as substantial fluctuations in oil prices. It has built-in flexibility and efficiency, and this should provide a more stable investment environment.

9

Economic Modeling/Auditing— Art and Science Part I

Discounted cash flow analysis is the workhorse of the upstream petroleum industry. Two characteristic features are the huge numbers involved and the enormous risks quantified and evaluated using the expected value approach outlined in Table 9–1. Unfortunately, the science of economic modeling and risk analysis typically outpaces the art. This is not good. It takes both.

Table 9–1 The Expected Value (EV) Formula

Expected Value = Reward x SP - Risk Capital x (1 - SP)
Where:
Risk Capital = Costs associated with testing a prospect. Typically consists of dry-hole costs, G&G costs, and possibly a signature bonus.
Reward = Present value of possible successful exploration efforts based upon discounted cash flow analysis of a hypothetical discovery typically discounted at (or close to) corporate cost of capital. [See Tables 9–2 and 9–3]
SP = Probability of success (likelihood of actually making a discovery as estimated by geotechnical personnel).
1 - SP = Probability of failure (likelihood of drilling a dry hole and losing the risk capital).
This is the basic equation of modern day risk analysis. The rule is: If expected value is positive, then the reward outweighs the risk. Companies try to choose investment opportunities that maximize expected value.

The artistic aspect of the exercise of cash flow analysis begins with the understanding that all economic models are flawed. Every model has its weaknesses—the challenge is to determine where these weaknesses lie, whether or not they are material, and/or if there are any fatal flaws. Too many managers make important big-dollar decisions based upon the results of economic models without knowing where the model's weaknesses lie nor how to locate them. Many problems are minor and not sufficiently material to justify all the effort required to rerun the economics, yet this can only be determined if the problems are noted and understood.

The example here is an economic model from an exploration scenario summarized in Table 9–2; however, the techniques can be used for evaluating farm-in/farm-out proposals, development feasibility studies, or production economics for acquisition or sale, etc. An economic (cash flow) model representing the exploration prospect summarized in Table 9–2 is shown in Table 9–3. The checklist in Table 9–4 provides guidelines for auditing the economic model. Estimates are made and compared with calculations from the model to check various aspects of the model.

Table 9–2 Case Study Parameters: Assumptions

Risk Model

Probability of success	15%
Risk capital	$4 MM
Reward side	Below (*see* Figure 9–1)

"Most Likely" Discovery Assumptions

100 MMBBL field (Assumed recoverable reserves if a discovery is made.)

26° API crude oil

GOR 800 cubic feet per barrel (gas/oil ratio)

Brent (North Sea) marker crude price at time of study—$22/BBL

600 feet of water

25 development wells drilled

3 development wells dry

110 feet of pay (Avg. reservoir thickness—productive section of the reservoir.)

6,000 acres productive area

Fiscal Terms

Type of system		Production sharing contract
Royalty		10%
Cost recovery limit		50%
Profit oil split	BOPD	Government share
	0 - 10,000	50%
	10,000 - 20,000	60%
	20,000 - 30,000	70%
	> 40,000	80%
Income tax rate	40%	
Depreciation rate	20%/year [for both tax and cost recovery]	

Table 9–3 Cash Flow Model for Assumed Most Likely 100 MMBBL Discovery

Year	Annual Oil Production (MMBBLS)	Oil Price ($/BBL)	Gross Revenues ($M)	Royalty 10% ($M)	Net Revenue ($M)	Capital Costs ($M)	Operating costs ($M)	Depreciation ($M)	Unrecovered Costs C/F ($M)	Cost Recovery ($M)
	A	B	C	D	E	F	G	H	I	J
1	0	$20				30,000			0	0
2	0	$20				40,000			0	0
3	500	$20	10,000	1,000	9,000	100,000	3,000	34,000	0	5,000
4	5,900	$20	118,000	11,800	106,200	60,000	15,800	46,000	32,000	59,000
5	9,312	$20	186,240	18,624	167,616	70,000	22,624	60,000	34,800	93,120
6	12,050	$20	241,000	24,100	216,900		28,100	60,000	24,304	112,404
7	10,750	$20	215,000	21,500	193,500		25,500	60,000		85,000
8	9,406	$20	188,120	18,812	169,308		22,812	26,000		48,812
9	8,230	$20	164,600	16,460	148,140		20,460	14,000		34,460
10	7,202	$20	144,040	14,404	129,636		18,404			18,404
11	6,301	$20	126,020	12,602	113,418		16,602			16,602
12	5,514	$20	110,280	11,028	99,252		15,028			15,028
13	4,825	$20	96,500	9,650	86,850		13,650			13,650
14	4,221	$20	84,420	8,442	75,978		12,442			12,442
15	3,694	$20	73,880	7,388	66,492		11,388			11,388
16	3,232	$20	64,640	6,464	58,176		10,464			10,464
17	2,828	$20	56,560	5,656	50,904		9,656			9,656
18	2,475	$20	49,500	4,950	44,550		8,950			8,950
19	2,165	$20	43,300	4,330	38,970		8,330			8,330
20	1,395		27,900	2,790	25,110		6,790			6,790
Total	100,000		2,000,000	200,000	1,800,000	300,000	270,000	300,000		570,000

Year	Total Profit Oil ($M)	Gvt. Profit Oil ($M)	Contractor Profit Oil ($M)	Tax Loss C/F ($M)	Taxable Income ($M)	Income Tax 40% ($M)	Contractor Cash Flow ($M)	
							Undiscounted	12.5% DCF
	K	L	M	N	P	Q	R	S
1	0	0	0	0	0	0	(30,000)	(28,284)
2	0	0	0	0	0	0	(40,000)	(33,522)
3	4,000	2,000	2,000	0	(30,000)	0	(96,000)	(71,514)
4	47,200	25,444	21,800	30,000	(11,000)	0	5,000	3,311
5	74,496	43,387	31,109	11,000	30,605	12,242	19,363	11,397
6	104,496	64,605	39,891		64,195	25,678	98,517	51,543
7	108,000	64,599	43,401		43,401	17,360	86,041	40,014
8	120,496	70,320	50,176		50,176	20,071	56,106	23,193
9	113,680	64,451	49,229		49,229	19,692	43,537	15,998
10	111,232	61,102	50,130		50,130	20,052	30,078	9,824
11	96,816	52,481	44,335		44,335	17,734	26,601	7,723
12	84,224	44,959	39,265		39,265	15,706	23,559	6,080
13	73,200	38,383	34,817		34,817	13,297	20,890	4,792
14	63,536	32,627	30,909		30,909	12,363	18,545	3,782
15	55,104	27,618	27,486		27,486	10,995	16,492	2,989
16	47,712	23,856	23,856		23,856	9,542	14,314	2,306
17	41,248	20,624	20,624		20,624	8,250	12,374	1,772
18	35,600	17,800	17,800		17,800	7,120	10,680	1,360
19	30,640	15,320	15,320		15,320	6,128	9,192	1,040
20	18,320	9,160	9,160		9,160	3,664	5,496	553
Total	1,230,000	678,692	551,308			220,523	330,785	54,357

Table 9–4 Cash Flow Audit Checklist

The Model		
1.	Government total profit oil share	
	At peak production	62%
	Average full-cycle	56%
2.	Government take	77%
	Contractor take	23%
3.	Effective royalty rate	30-35%
4.	Savings index	26¢
5.	Entitlement	56%
6.	Spot checks	various
Price and Cost Assumptions		
7.	Initial oil price	$20/BBL
8.	Capital cost per unit	$3/BBL
9.	Capital costs per BOPD	$9,090/BOPD
10.	Capital costs as a percentage of gross revenues	15%
11.	Total costs as a percentage of gross revenues	28.5%
12.	Operating costs (peak year)/total costs	9.4%
13.	Operating costs (early years) ($/BBL)	$2.50±/BBL
14.	Operating costs (full cycle)($/BBL)	$2.70/BBL
Technical Aspects		
15.	Peak production/total reserves	12%
16.	Decline rate	12.5%
17.	Well spacing (acres per well)	270 acres
18.	Initial production rate per well (BOPD)	1,500
19.	BOPD per foot of pay (reservoir thickness)	14 BOPD/ft
20.	Development drilling success ratio	88%

The risk/reward equation referred to as the EV or EMV model summarized in Table 9–1 shows where discounted cash flow values (from economic models like that in Table 9–3) are used.

The EV formula, whether it is used directly or indirectly (gut feel or instinct), provides the basis for billions of dollars of exploration investments. It is normally more complex with the common practice in the industry of using multiple outcomes (at least three) on the *reward side* of the equation. This is illustrated in the decision tree in Figure 9–1. The focus of this chapter is on how the *reward side* values are derived. The cash flow model summarized in Table 9–3 represents the *most likely* outcome (100 MMBBLS) if a discovery is made. This model is subjected to examination. Certainly the same procedure would be applied to other models representing possible *maximum* and *minimum* outcomes.

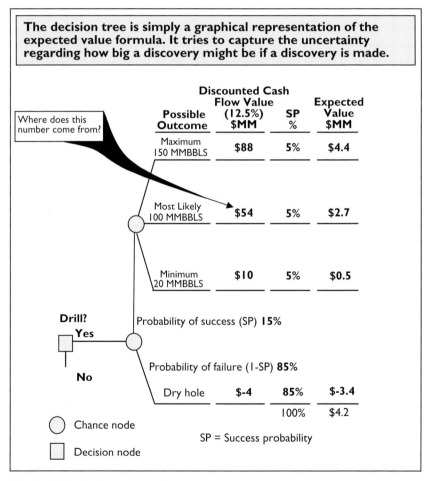

The decision tree is simply a graphical representation of the expected value formula. It tries to capture the uncertainty regarding how big a discovery might be if a discovery is made.

Where does this number come from?	Possible Outcome	Discounted Cash Flow Value (12.5%) $MM	SP %	Expected Value $MM
	Maximum 150 MMBBLS	$88	5%	$4.4
	Most Likely 100 MMBBLS	$54	5%	$2.7
	Minimum 20 MMBBLS	$10	5%	$0.5
	Probability of success (SP) 15%			
Drill? Yes / No	Probability of failure (1-SP) 85%			
	Dry hole	$-4	85%	$-3.4
			100%	$4.2

SP = Success probability

○ Chance node
☐ Decision node

Fig. 9–1 Multiple outcome decision tree

Discussion

Each measure in Table 9–4 provides its own particular insight into the veracity of the model. In addition, there are a few measures provided to show where important, often-used figures come from. Some measures take on greater meaning when viewed in the context of others—as well as an understanding of the area, region, or play. Some of these measures are more useful and powerful than others. Some are rather obscure. Yet in the context of other measures, and with increased usage, they take on added value.

The objective here is to help analysts, auditors, and managers improve their analytical skills and provide guidelines that will give them more confidence wending their way through an economic/cash flow model.

The Model

With so many cash flow programs, spreadsheets, black-boxes, and modeling techniques around these days, it is somewhat dangerous to simply assume that contract terms have been modeled correctly. There are a number of ways to check the veracity of a particular model. While there are many things that can be done to audit the modeling itself, some of the key techniques are shown here. Because the PSC has a sliding scale, this aspect needs to be examined first.

Government Total Profit Oil Share (56%)

The example fiscal system has a sliding scale profit oil split similar to many found in PSCs around the world. Roughly 80% of the PSCs worldwide have sliding scales. In order to evaluate various other aspects of the model, first take a closer look at the sliding scale.

A quick estimate of what the weighted average profit oil split (or any other parameter for that matter) would be over the life of a field requires only three basic steps:

1. Estimate what the split would be in a peak year of production. In the cash flow model, production peaks at 12,050 MBBLS in Year 6. This comes to approximately 33,000 BOPD. At 33,000 BOPD, the profit oil split is roughly 62/38% in favor of the government, as shown in Table 9–5.

2. Estimate the split at the end of the life of the field.

3. Take the average of the two.

Table 9–5 Government Profit Oil Share Estimate (Full Cycle)

Step 1						
Tranche	Govt. P/O Split	Govt. Share				
1st	10,000/33,000	×	50%	=	15.15%	
2nd	10,000/33,000	×	60%	=	18.18%	
3rd	10,000/33,000	×	70%	=	21.21%	
4th	3,000/33,000	×	80%	=	7.27%	
Government share peak year					61.81% [year 6]	
Step 2						
Government share last year					50% [year 20]	
Step 3						
Average	=	$\dfrac{61.81 + 50\%}{2}$		=	56% [Total, full cycle]	

Thus the estimated full-cycle profit oil split is 56/44% in favor of the government. Using this technique of comparing the peak and ending splits will usually provide an accurate estimate (*see* Fig. 9–2). If the peak production had by far exceeded the highest tranche, then the estimate would likely have been low. For example, the highest tranche is >40,000 BOPD, and if production had reached, say, 80,000 BOPD, this technique would typically underestimate the government share of profit-oil (full-cycle) and overestimate the company share.

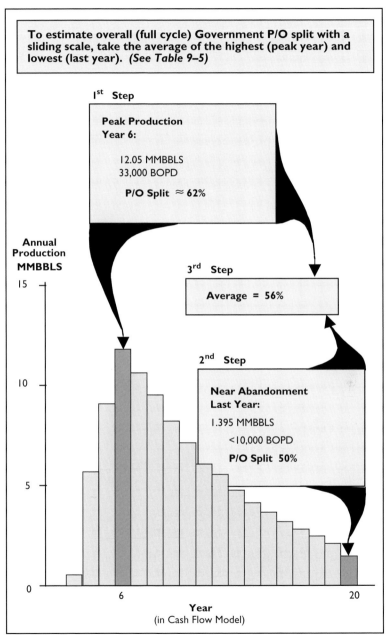

Fig. 9–2 Government profit oil share estimate (full cycle)

EXAMPLE: "MOST LIKELY" 100 MMBBL FIELD PRODUCTION PROFILE

To estimate overall (full cycle) profit oil split with a sliding scale, take the average of the highest (peak year) and lowest (last year) [*See* Table 9–5]. The actual full cycle division of profit oil from the cash flow model (calculated not estimated) is shown in Table 9–6.

Table 9–6 Government Profit Oil Share Calculation ($M)

Year 6 (peak production)			
Government profit oil			$64,605
Total profit oil			$104,496
Govt. Share	=	$\dfrac{\$64,605}{\$104,496}$ =	61.8%
Full Cycle			
Government Profit Oil			$678,692
Total Profit Oil			$1,230,000
Govt. Share	=	$\dfrac{\$678,692}{\$1,230,000}$ =	55.18%

Total Govt. Profit Oil Share 55.18% (calculation) vs. 56% (estimate)

One reason for this modest difference is that profit oil splits are calculated on the basis of gross production, yet applied to profit oil only. Profit oil, as a percentage of total production in the early *saturated* years when cost recovery is at the limit, is only 40% of production in this fiscal system. These are typically the years during which the government share of profit oil is greatest. Later after payout when production rates are lower, profit oil may represent more than 70% of production. The suggested three-step estimate does not take this into account. However, the estimate will often closely mirror detailed year-by-year estimates.

Government Take (77%)

In many countries, the fiscal/contract terms are so well known that this simple calculation of government take could indicate whether or not there might perhaps be a problem with the economic model itself. For example, if this project were in Malaysia under the late 1990s vintage contract, then the government take would be expected to be on the order of 83% or so, not below 80%.

The example here uses a fairly common system, and a quick back-of-the-envelope estimate indicates that the cash flow model is likely on track. It yields a government take of 77.4% in Table 9–7. Approximately 85% of all fiscal systems worldwide can be quickly checked this way. Those fairly rare systems—with depletion allowances, R factors, price cap formulas, ROR features, and excess cost oil provisions—can get a bit complicated.

Table 9–7 Government Take Estimate

100%	Gross revenues (full cycle)
-10	Royalty rate
90	Net revenues
-30	Total costs
60	Total profit oil
-33.6	Govt. share profit oil (56%, est.)
26.4	Contractor share profit oil
-10.6	Tax 40%
15.8	Company cash flow
Contractor take =	15.8/(100 - 30) = 22.6%
Government take =	(10 + 33.6 + 10.6)/(100 - 30) = **77.4%**

Government Take: 77.4% (estimate) vs. 76.9% (calculation)

So why was there a difference between the calculations from the model and the estimate? And is this a significant difference? The back-of-the-envelope estimate (Table 9–7) that yields a take estimate of 77.4% is based on a slightly different cost assumption. The costs as a percentage of gross revenues used in Table 9–7 equal 30% vs. 28.5% in the model. This would make only a slight difference because the royalty is not that large. The biggest difference arises from the profit oil split estimate of 56% (Table 9–5 and Figure 9–2) vs. the 55.18% profit oil split calculation from the model. The cash flow model provides a more detailed year-by-year estimate. However, it appears that the difference is quite small. Based upon the slight difference between the cash flow take of 76.9% from the cash flow model (*see* Table 9–8) and the quick-look estimate (77.4%), it is likely that there are no big errors in the model. There are other things to check, of course.

Table 9–8 Government Take Calculation

Total profits	=	Gross revenues – total costs
Total profits	=	$2,000,000M - $300,000M – $270,000M
	=	$1,430,000M
Govt. take	=	Govt. share of profits/Total profits
Govt. share	=	$200,000M + $678,692M + $220,523M
	=	$1,099,215M
Government take	=	$1,099,215M/$1,430,000M = **76.9%**

ERR (30–35%)

With any PSC that has a cost recovery limit, the government will be guaranteed a share of production in each accounting period by virtue of the combination of the limit and the subsequent profit oil split. This creates much of the effect of a royalty. For PSCs—like the example system here where there is also a royalty—the effect is magnified. The combined effect of a royalty and a cost recovery limit is referred to as the ERR. It represents the minimum share of revenues the government might expect through royalty payments and profit oil in any given accounting period. With production-based sliding scales, this minimum guarantee will change from accounting period to accounting period as production rates change.

An estimate is provided in Table 9–9. Here it is assumed that the cost pool by far outweighs the available revenues and the system is *saturated*, i.e., at the limit. Furthermore, with sufficient deductions, the company would be in the position of paying no taxes. Thus, the ERR will range from 30 to 34.8%, depending upon the profit oil split in any given accounting period because of variations in production levels.

Table 9–9 Effective Royalty Rate Estimate

Minimum	Maximum	
100%	100%	**Gross revenues**
-10	-10%	Royalty rate
90	90	Net revenues
-50	-50	Total cost recovery [saturated]
40	40	Total profit oil
-20	24.8	Govt. share profit oil (50%–62%)
20	15.2	Contractor share profit oil
-0	-0	Tax (40%)
30%	34.8%	**Effective royalty rate [royalty + profit oil]**

Notice that in the early years of production in the cash flow model there is a cost recovery C/F. Also there is no taxable income in the first two years of production. In these two years, the ERR can be taken from the cash flow model.

An example from Year 4 is shown in Table 9–10. In this particular year, the company is in a no-tax-paying position and the government receives only 31.5% of production as shown and discussed as follows:

$$(\$11,800 + \$25,400)/\$118,000 = 31.5\%$$

Table 9–10 Effective Royalty Rate—Year 4 ($M)

$118,000	Gross revenues from the model
-11,800	Royalty
106,200	Net revenues
-59,000	Total cost recovery
47,200	Total profit oil
-25,400	Govt. share profit oil* (53.8%)
21,800	Contractor share profit oil
-0	Tax
$37,200	Govt. share of revenues [$11,800 + 25,400]
31.5%	Effective royalty rate* [$11,800 + 25,400]/$118,000

*Year 4 "off peak" production rate is 16,160 BOPD.

In Year 4, the gross revenues are $118,000 M. Of this, the government receives $11,800 M in royalties and $25,400 M in profit oil and no taxes. This kind of situation where a company can be in a no-tax-paying position can happen under a variety of circumstances: in the early stages of even a profitable field, in the latter stages of production for all fields, and during much of the life of marginal and sub-marginal fields.

Savings Index (26.4¢)

There is much discussion these days about the *mutuality* or *alignment* of interests between host governments and IOCs as an important objective in fiscal/contract design, and most of the context of this design concept deals with creating incentives for cost savings. To a large extent, this can be measured.

Typically (not always) for any given fiscal system, an oil company (and the government) will benefit from a reduction in costs—either capital or operating. And the degree to which the company will benefit depends upon the profits-based fiscal elements. For example, in this fiscal system, there is a profit oil split in favor of the government on the order of, say 56%, and a tax of 40%. (This is dealt with in more detail later). The combination of these two (profits-based) levies will yield an effective tax rate of 73.6%. Thus if a company manages to save one dollar, it will be able to keep 26.4% of that dollar as shown in Table 9–11. The savings *index* is 26.4% or 26.4¢ on the dollar (saved).

Table 9–11 Savings Index Estimate

$1.00	Saved (yields an additional $1 of profit oil
.56	Govt. share of profit oil (56%)
$0.44	Additional taxable income from $1 saved
-0.176	Tax (40%)
-0.264	Contractor share of $1 savings
26.4¢	on the dollar (undiscounted)

In terms of present value, the benefit of a dollar saved will be different than the index. From an exploration point of view, reducing capital expenditures by a dollar may increase company cash flow discounted at 12.5% by more than 26.4%. The same is not true of operating expenses. Saving a dollar of operating expenses in the Table 9–3 cash flow model might improve contractor discounted cash flow by only 10¢ or so. However, in any given year if management reduces operating costs the impact will be closer to the index—26.4¢.

Entitlement Index (56%)

Booking barrels is much more common these days than it was 10 years ago. Under most systems, companies will *book* the equivalent of their *entitlement* barrels net to their working interest share of proved reserves only. Under a PSC, *contractor entitlement* consists of cost oil plus profit oil. *Government entitlement* consists of royalty oil and profit oil. In a typical model, cost oil and profit oil have been converted to dollars so a simple calculation converts them back to percentage entitlement and then barrels as shown in Table 9–12. The Table 9–7 estimate of government take provides the components of contractor entitlement: cost oil 30% and profit oil 26.4%, which yields an entitlement estimate of 56.4%.

Table 9–12   Entitlement Calculation ($M)

$570,000	Company cost oil ($)
$551,308	Company profit oil ($)
$1,121,308	**Company entitlement ($)**
56%	**Company entitlement (%)**
	[$1,121,308/$2,000,000]
168,000 MBBLs	**Company entitlement**

This "lifting entitlement" would, however, not correspond to the reserves the company would be able to "book" for Securities and

Exchange Commission (SEC) purposes. Even in the event of a 100 MMBBL discovery, it would be quite a long time before all 100 MMBBLs would qualify as *proved* reserves. Thus—depending upon the proved reserve estimate—the company would likely *book* roughly 56% of those barrels.

Spot Checks

There is already evidence to indicate that the model is working as it should, but the indicators used so far are not sufficient. There are many other ways to check the model, and some are shown here.

C/R Limit

A quick check in the early years of production will often show if the model is honoring the 50% cost recovery limit. It depends upon whether or not the system is *saturated* or whether or not the limit is tested. In this case (in Year 4), there are unrecovered costs carried forward. Gross revenues are $118,000 M, and total cost recovery that year is projected at half of that—$59,000 M—as it should be.

40% Tax

The tax rate is supposed to be 40%, and this can be checked in individual years against taxable income. It can also be checked against contractor profit oil. In any given accounting period, the company share of profit oil will not be the tax base; but usually over the life of a field it will average out (*see* Table 9–7). Therefore, to check the model, income tax paid comes to $220,523 M or 40% of the company share of profit oil: $551,308. An additional check is provided in Table 9–13 with an inspection of the tax base calculation in Year 5.

Table 9–13 Income Tax and Cash Flow Calculations—Year 5 ($M)

Taxable Income =	Gross revenues - Royalties - Depreciation - Operating costs (OPEX) - Government profit oil - Tax loss carry forward
Taxable Income =	$186,240 - 18,624 - 60,000 - 22,624 - 43,387 -11,000
=	**$30,605 Check**
Net Cash Flow = (after-tax)	Gross revenues - Royalties - Capital costs - Operating costs - Government profit oil - Taxes
Net Cash Flow = (after-tax)	$186,240 - 18,624 - 70,000 - 22,624 - 43,387 - 12,242 [.4 x $30,605 above]
=	**$19,363 Check**

Company Cash Flow

Table 9–13 also provides a spot check (Year 5) of the model by testing the calculation of company cash flow. This provides additional assurance that the modeling has been constructed correctly. Assuming the model is correct, the next step is to inspect various assumptions.

CONCLUSION

It takes a lot of experience to have a feel for whether or not certain assumptions are in tune and balanced. Of course, it also requires a knowledge of how and where to check—that is what this book is all about. This example audit exercise has purposely been generic and non site-specific. In practice, this exercise would be conducted within the context of known conditions, such as offset fields, experience in the region, basin, and play, and so forth. This kind of information in conjunction with the methodology outlined here creates powerful tools. Part Two of this exercise (Chapter 10) will extend the audit exercise to include examination of the assumptions regarding capital costs, operating costs, decline rates, and other factors that go into an economic model.

10

Economic Modeling/Auditing— Art and Science Part II

The artistic aspect of the exercise of cash flow analysis begins with the understanding that all economic models are flawed. Every model has its weaknesses—the challenge is to determine where these weaknesses lie, whether or not they are material, and/or if there are any fatal flaws. Too many managers make important big-dollar decisions based upon the results of economic models without knowing where the model's weaknesses lie nor how to locate them. The example audit exercise in Chapter 9 is purposely generic and non site-specific. This chapter extends the audit exercise to include an examination of some of the key price, cost, and technical assumptions.

Initial Oil Price ($20/BBL)

It was assumed that at the time of the analysis, North Sea marker Brent Crude was trading at $22/BBL. The oil price in the model is $2/BBL lower than the price of the lighter Brent crude. Brent is a well-known North Sea marker crude blended from the fields producing into the Brent and Ninian pipelines. Because the mix of crudes has changed over the years, the API gravity has changed slightly but not dramatically. Brent crude is approximately 39° API. The question, of course, is whether or not the price adjustment assumed in the economic model is sufficient. Heavier crudes are not as valuable as lighter crudes (like Brent), and the price adjustment can range from as low as 1.5% per degree API to as much as 3%. Assuming a price adjustment of 1.5%/degree the 26° API crude would sell for $17.70/BBL.

Brent $22.00/BBL
Adjustment   - 4.03/BBL [13° API x 1.5%/degree = 9.5% adj.]

Adjusted price
    estimate $17.70/BBL

The adjustment could have been greater, of course. The generic relationship is shown in Figure 10–1.

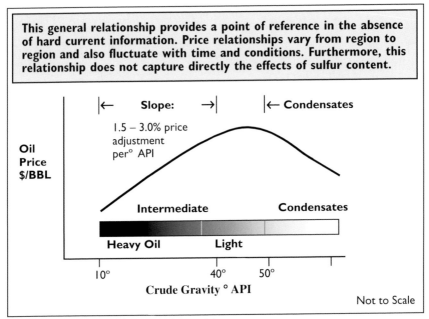

Fig. 10–1 *General oil price relationship*

Capital Cost per Unit ($3/BBL)

$300,000 M/100,000 MBBLS = $3.00/BBL

Capital costs are an extremely important aspect of project economics. The main categories of cost associated with the upstream petroleum industry are: exploration, development, operating, abandonment, and financing costs (cost of capital).

As far as economic sensitivity is concerned, exploration risk capital is about 10 times as important as development capital, and development capital costs typically outweigh operating costs by a wide margin as well. Development costs typically consist of:

- Drilling costs
- Production/processing facilities
- The transportation function.

In this case, $3/BBL could be a reasonable number. It is, however, slightly low by world standards, particularly for 600 feet of water. There is not sufficient information in this exercise to make comparisons. However, in the real world, analysts would have a feel for whether or not this is a reasonable number in a particular area.

Exploration Well Costs

The cost of drilling an exploratory well is a useful index. Exploration drilling costs often constitute the lion's share of the risk capital associated with an exploration venture. Knowing how much it costs to drill an exploratory well will also provide some insight into subsequent development drilling costs should there be a discovery.

Drilling costs typically can represent 25–50% of the total costs associated with a development. Production facilities, of course, become of greater and greater importance the more remote the location and the deeper the water. Furthermore, production facility costs are quite sensitive to GOR. The facilities required to handle large volumes of gas can be quite expensive. The natural question of course is: *What constitutes a high GOR?*

GOR is typically measured in terms of cubic feet per barrel. The famous engineer J. J. Arps had a rule-of-thumb: typically GOR for black oil is equal to reservoir depth divided by 10. For example, if oil was found at a reservoir depth of 7000 ft, then the amount of gas in solution would be equal to 700 cu ft per barrel (7000/10). This would not require substantial gas-handling facilities by world standards. However, some crudes can have a GOR of 3000 to 9000 cu ft per barrel or more. In the case of the 100 MMBBL field example in the peak year of production, 3000 to 9000 cu ft per barrel would require facilities capable of handling 100 to 300 MMCFD. Just 100 MMCFD alone would feedstock a 600 mw gas-fired power plant.

Capital Costs per BOPD ($9,090/BOPD)

$300,000 M/33,000 BOPD = $9,090/BOPD

This statistic is based on total capital costs divided by peak daily production. Peak production is projected in Year 6 when 12,050,000 barrels are modeled (around 33,000 BOPD). This is an interesting statistic—perhaps more interesting than useful as far as this particular audit exercise is concerned. However, it is a statistic used in a variety of circumstances and for that reason is included here to show where these numbers originate.

Often, when macro-economists discuss the capital cost requirements to meet world demand growth (for crude oil) they will state that OPEC— particularly the big four Gulf states, Iran, Iraq, Kuwait, and Saudi Arabia— will need to add another 15 MMBOPD of capacity in the next 7 to 8 years. Capital cost requirements are estimated to be on the order of $60 billion. This equates to $7.5 billion per year for the next 8 years for upstream capacity only. This is based upon the assumption of $4000/BOPD of capacity. Refining capacity expansion is often rated in terms of $12,000 per barrel per day of capacity, and this would include both distillation capacity as well as typical upgrading units such as cracking and reforming and such.

Capital Costs as a Percentage of Gross Revenues (15%)

$300,000 M/$2,000,000 M = 15%

Total capital costs divided by gross revenues is 15%. By world standards with $20/BBL (for a Brent-quality crude) this is a fairly normal percentage. It is no surprise that as capital costs increase, project economics deteriorate. The point at which costs become too high is usually very close to where capital costs as a percentage of gross revenues (15%) approach contractor take.

Contractor take in this system is 23% so it is not surprising that net present value (NPV) at a 12.5% discount rate is positive. Had government take been greater than 85%, it is likely that the present value (12.5%) would have been negative.

Total Costs as a Percentage of Gross Revenues (28.5%)

($300,000 M + $270,000 M)/$2,000,000 M = 28.5%

Total costs (including both Capex and Opex) divided by total revenues under ordinary conditions (if there is such a thing) are often around 35%. In this model, the ratio was less than 30%. This is not unusual. Governments are extremely sensitive to costs. In their view, if sufficient revenues are generated all costs borne by the oil companies are reimbursed out of revenues generated from the government's mineral resources. Every dollar of additional cost reduces government profits. The same is true for oil companies.

Operating Costs (Peak Year)/Total Capital Costs (9.4%)

$28,100 M/$300,000 M = 9.4%

This is a fairly obscure but useful statistic. For conventional developments, the range is often from 3% to 8%. In the Gulf of

Mexico shelf, the relationship between annual operating costs and total Capex is often from 3% to 5%. In the U.K. North Sea, the range might be more like 6% to 8%. World average is probably close to 5%. However, for deepwater non-conventional developments with substantial floating elements for production, storage, and off-loading, the ratio can approach 20% or more.

Operating Costs (Early Years) ($2.50±/bbl)

$15,800,000 M/5,900 MBBLS = $2.67/BBL   [Year 3]
$22,624,000 M/9,312 MBBLS = $2.43/BBL   [Year 4]

Operating costs in the early years of production are assumed to be roughly $2.50/BBL. This may seem a bit low, depending on the region and the particular situation. Average operating costs worldwide are probably higher by about $1/BBL. It is certainly possible for costs to be this low, but there is not enough information in this exercise to say one way or another. Analysts would likely know for a particular area whether this number was relatively high or low.

Operating Costs (Full-Cycle) ($2.70/bbl)

$270,000 M/100,000 MBBLS = $2.70/BBL

In any given region or situation, there is a likely level of operating costs that would be considered realistic or reasonable under a given set of conditions. Full-cycle operating costs with most models are typically higher than operating costs per unit in the early years of production. Sometimes when analysts or management quote operating costs, they include depreciation. Furthermore, many countries define as operating costs expensed costs and intangible costs that are not required to be capitalized (i.e., these costs are expensed not amortized). In order to be comfortable with operating costs on the order of $2.70/BBL, there should be a good healthy economy-of-scale and no cruel and unusual conditions that might require higher costs. A 100,000 MBBL field is not necessarily large, but it may be adequate to provide sufficient economies to justify lower-than-average operating costs.

The key factors that influence both capital and operating costs include:
- Water depths or terrain
- Climate: weather windows, wave conditions, or spring break-up
- Infrastructure: roads, rail, port facilities, airports, or communications
- Distance from supply points for goods and services
- Distance to market
- Reservoir depth
- Rock type
- Petrophysical parameters
- Reservoir pressure gradient
- GOR
- Fluid properties: paraffin content (pour point)
- Political conditions/risks

TECHNICAL ASPECTS AND ASSUMPTIONS

The basic unit of production in the upstream end of the industry can be viewed as either a well or a field depending upon the situation. In this case both are considered. The petroleum industry is highly technical. The measures here only scratch the surface, but they do provide some insight into the dynamics of the model. Furthermore, checking these aspects of the model provides quick indications of particular areas for further inspection.

Peak production/total reserves (P/R) (12%)

12,050 MBBLS/100,000 MBBLS = 12.05%

In Year 6 of the cash flow model, production peaks at 12,050 MBBLS. This represents 12% of the total reserves. This production/reserve

(P/R) ratio is a useful and direct measure of the rate of production. Typically, field developments are designed in such a way that roughly 10% or so of the recoverable reserves are produced in a peak year of production. However, higher rates can be found. Indonesia is fairly famous for high P/R ratio—on the order of 20 to 25%. Often the production decline rate coming off plateau production will be close to or greater than the P/R ratio. The rate of production can have a huge impact on project economics. Therefore, it is important to ensure that if a production profile has a particularly high (or low) P/R ratio that there is adequate justification.

Decline Rate (12.5%)

Typically, production decline rates will be equal to or greater than the P/R ratio. In this case, there is little information provided by this statistic, but it was considered important to illustrate how this can be observed fairly quickly. Take Years 10 and 16, for instance. Year 10 from the production profile has 7202 MBBLS of production. The previous year was 8230. This represents a 12.5% decline. The same rate is found in subsequent years.

$$\textbf{Rate of change} \quad = \quad \frac{7,202 \text{ MBBLs [Year 10]}}{8,232 \text{ MBBLs [Year 9]}} \quad = 87.5\%$$

$$\textbf{Decline rate} \quad = \quad (1 - 87.5\%) \quad = \textbf{12.5\%}$$

Well Spacing (270 Acres)

6,000 Acres/22 Wells = 270 Acres/Well

One of the important companion statistics to the P/R ratio is well spacing. An oilfield can be produced either slowly or very quickly, and much of this will be reflected in the P/R ratio. It depends primarily on the number of wells a company is willing to drill. Up to a certain point, there is an advantage to drilling more wells, and beyond that point, there are diminishing returns. The objective is to maximize present value.

There is not sufficient information to determine if the spacing is appropriate, especially without a map and other information. However, it is likely that it would be difficult to produce as much as 12% of the reserves in one year from a given reservoir on such a large spacing if all the wells are vertical. In the early days of the North Sea, developments were typically initiated with 200–240 acre spacing. Later, additional in-fill wells were drilled.

Now, more horizontal wells are being drilled. Typically, horizontal wells extend at least 3000 feet horizontally through the reservoir. Much less than that is a waste, and beyond that the relative benefits typically diminish substantially. With a 3000 foot horizontal leg, a well will drain approximately twice as much as a vertical well on a 200-acre spacing. This is illustrated in Figure 10–2. If all the wells in this 100,000 MBBL development were horizontal, then it may likely produce faster than the 12% P/R in the model.

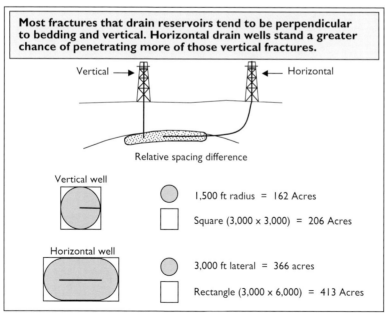

Fig. 10–2 Vertical vs. horizontal wells, relative spacing difference

Initial Production Rate per Well (1500 BOPD)

33,000 BOPD / 22 Wells = 1500 BOPD / Well

The initial production rate per well is an extremely important parameter. From the limited information provided with this model, only an estimate can be made. Supposedly 22 wells are assumed to be productive. However, from the model, it cannot be determined just how many are actually producing in the early years. Therefore, it will be assumed that by Year 6 of the model all wells are producing. Yet some of the wells will already have been producing for more than 3 years by that time. Thus, this estimate will be somewhat low. However, this is an accuracy vs. precision issue, and a simple estimate is sufficient.

The field is projected to be producing at 33,000 BOPD by Year 6 (Year 4 of production). Divided among 22 wells, this yields an initial average rate of around 1500 BOPD/well. Of course, the question is: *Is this reasonable?* It depends on many things.

Vertical vs. horizontal wells, relative spacing difference

One common mistake is to use reported test rates from an area in an exploration or development model. For development feasibility economics, sometimes test rates from the discovery and appraisal wells are used. This is often not appropriate. Reported test rates are usually the result of *combined* flow rates that may include separate drillstem test results from numerous reservoir intervals up and down the hole.

Table 10–1 shows reported discovery well test rates worldwide for the years 1996–1998. The average test rate for an oil discovery during this period was around 5000 BOPD. It is likely that the average production rate per well during the first full year of production for development wells associated with these discoveries would be half this much.

Table 10–1 Reported International Discoveries 1996–1998

Well Test Rates

In any given province, the delivery rate per well is important, but it is of particular significance for deepwater developments. Of the discoveries reported below, approximately half were onshore and half offshore.

Year	Reported *Oil* Discoveries	Combined Flow Rate on Test (BOPD)		
		Lower Quartile	Average	Upper Quartile
1996	46	940	4,600	9,900
1997	55	740	6,900	10,970
1998	84	350	4,015	11,070
	185	613	5,018	10,750

Year	Reported *Gas* Discoveries	Combined Flow Rate on Test (MMCFD)		
		Lower Quartile	Average	Upper Quartile
1996	57	4	22	50
1997	28	3	18	43
1998	53	2	22	62
	138	3	21	53

Summarized from: *AAPG Explorer*, Jan., 1996, 1997, 1998. Major discoveries compiled by Petroconsultants.

BOPD per Foot of Pay (14)

1500 BOPD/110 feet = 14 BOPD/ft

This is an extremely rare statistical measure. In fact, it should be used with extreme caution. For one thing, this productivity index captures the effects of only one (pay thickness) of four main parameters that influence deliverability. Deliverability is also directly proportional to permeability and drawdown (pressure differential in the well bore), and inversely proportional to fluid viscosity.

However, for those familiar with both strengths and weaknesses of this yardstick, it does provide some insight. A productivity index of 14 BOPD/ft is not high. In fact, 10–20 is about average. Anything higher than 40 better have a good reason. Some wells produce at rates on the order of 60 to 80 BOPD/ft, but these typically occur because of dramatically reduced bottomhole pressure. Submersible pumps will do this.

Development Drilling Success Ratio (88%)

22 Productive Wells / 25 Development Wells Drilled = 88%

Even development wells can come up dry. In some areas, in fact, the ratio can be quite high—on the order of 20%. Furthermore, there are worse things than a dry hole. A blowout, of course, would qualify, but perhaps more common than that is the kind of drilling where a completed well yields insufficient production to justify even the completion costs—let alone the dry hole costs.

CONCLUSION

It takes a lot of experience to have a feel for whether or not certain assumptions are in tune and balanced. Of course, it also requires a knowledge of how and where to check. Many problems are not sufficiently material to justify all the effort required to rerun the economics, yet this can only be determined if the problems are noted and understood. In practice, this exercise would be conducted within the context of known conditions, such as offset fields; experience in the region, basin, and play; and so forth. This kind of information in conjunction with the methodology outlined here creates powerful tools.

11

Finger on the Pulse—Phuket 2001

Periodically, governments of petroleum-producing nations and oil companies gather for *roundtable* discussions of petroleum fiscal systems and PSCs. Typically about 15 governments are represented and about a dozen oil companies. The one held in late June 2001 in Phuket, Thailand provided an excellent opportunity to get a finger on the pulse of what is happening around the world. Below are some highlights of that 2½-day event.

THE CONCEPT OF GREED

There is still a prevailing belief among some oil company personnel that most governments are greedy and that terms are too tough in many countries. This belief creates a poisonous attitude that is often impossible to conceal. It always surfaces in a setting like this and typically elicits responses similar to the one made by a representative of Pertamina, the Indonesian NOC: "We signed nearly 20 contracts last year." Indonesia is famous for tough PSC terms and receives numerous accusations of greediness, but complaints aside, the oil companies are still signing contracts. Perhaps another perspective might be helpful, the relationship between oil companies and the service industry.

Figure 11–1 illustrates a situation where a company invites tenders for goods and services. In this example, three service companies respond, and assuming they are all technically competent and capable of providing these goods and services then the choice is a simple one. The lowest bid is the winning bid. This is the marketplace at work—free enterprise. It is this mechanism that ensures that the government too will ultimately pay less when an oil company is reimbursed for costs incurred.

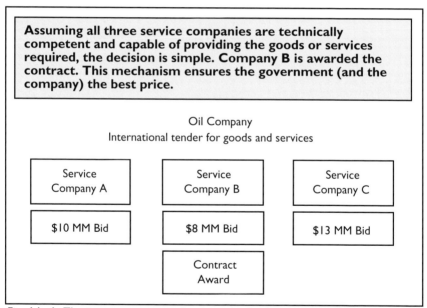

Fig. 11–1 *The procurement mechanism: this is the essence of the marketplace at work*

Governments do the same thing when tendering licenses for exploration rights. They award licenses to the highest bidder assuming the bidders are technically and financially qualified. It is difficult to find any difference between oil companies procuring goods and services and governments awarding exploration rights. Oil company personnel often argue that there is a huge difference between these two examples because of the cold, harsh specter of risk associated with exploration— as if service companies do not experience risk. I do not agree.

In the early 1990s the day rates for third-generation North Sea semisubmersible drilling rigs was down to around $30,000/day. This is a good barometer for many other services associated with international exploration. Service companies were suffering. Considering it takes $200 to 250 million dollars to build these rigs, such an investment would justify a day rate about six times higher. A common and fairly useful rule-of-thumb is to figure about a $700 day rate for every million in construction cost. If the drilling companies had predicted the downturn in the early 1990s, they would have built fewer semis, but they didn't know. They had to take their chances. Some win. Some lose.

Andrew Jupiter, the Permanent Secretary from Trinidad and Tobago, explained deepwater licensing efforts and fiscal design in his country. At one point, they launched an effort to design and structure terms to accommodate their deepwater environment and conditions. In fact a special group was organized for this project, but the group was ultimately disbanded because oil companies were making proposals much more robust than expected. The government essentially decided to evaluate all the proposals and take the best ones. This is how the system works.

That is not to say that *greed* does not exist. One of the delegates proposed an interesting example. A country awarded offshore exploration rights in the early 1980s for signature bonuses totaling $156 MM (approximately). Unfortunately, the government ultimately would not allow the companies to actually fulfill their drilling obligations: the government would not allow the companies to either drill *or* recoup the bonuses. More than 19 years later, the resulting dispute finally ended up before the U.S. Supreme Court, and the companies won this final appeal—but not before they lost an important round in a federal appeals court. This story is oversimplified, but, quite obviously, something is wrong here.

The competition is heavy for exploration acreage and projects around the world, and this must be frustrating for many companies. A few years ago, I worked with a government that received an extremely aggressive bid for a block from a U.S. oil company. There was little choice but to award the exploration rights to that company, but we were shocked at the terms they offered. At Phuket, I discussed progress with my client. The company is having a difficult time finding partners. The terms are too tough. The terms are indeed tough, but not because of a greedy government.

Government Take

While government take statistics provide important information and are still discussed extensively in industry literature, confusion persists. Even in these conferences populated by experienced people, the term *government take* has many different interpretations. Adding to the confusion is the plethora of other terms used to mean essentially the same thing: *state take*, *tax take*, *fiscal take*, and *government take of the profit*. Government take as a statistic has strengths and weaknesses.

How Does Indonesia Get Away With Such Tough Terms?

Indonesians have had an excellent feel for what the market can bear for many years. And not all Indonesian terms are necessarily tough. Furthermore, the *toughness* of terms generally measured by government take does not capture everything. It does not ordinarily take into account the relatively generous combination of ringfencing and *relinquishment* that adds such an interesting dynamic to Indonesia licensing agreements.

In Indonesia, the license area as a whole is typically ringfenced. There are no harsh internal ringfences that segregate a discovery from the rest of the acreage. If a discovery is made within the license area, then subsequent exploration costs can be cost recovered and tax deducted. With fairly large licenses and liberal relinquishment provisions if a company makes a discovery, then subsequent efforts carry almost no risk by most standards around the world. (There is potential for an internal ringfence between oil and gas field developments. Countries usually must do this if there are different fiscal terms for oil and gas.)

The relinquishment provisions are also fairly liberal. Typical Indonesian relinquishment requires the oil company to relinquish 25% of the original contract area at the end of the first and second exploration phases. Thus, if the contractor has made a discovery, it has 50% of the original contract area remaining. Many other countries would require relinquishment of all remaining acreage other than development areas or discoveries at the end of the exploration term. Furthermore they would require all development areas to be treated separately—that is, they would erect a ringfence around them. Indonesian ringfencing/relinquishment is generous by world standards and this does not get captured in the various *take* statistics.

KEEPING COSTS DOWN

Andrew Jupiter, permanent secretary from Trinidad, was asked by a delegate from an NOC: "Do your deepwater contracts provide incentives for oil companies to keep costs down?" This is a frequently asked question. Mr. Jupiter's answer was relatively simple: If exploration efforts are unsuccessful, the companies recover nothing, so the incentive to spend efficiently is great. If a discovery is made and profits are generated, then for every dollar saved, the companies will keep nearly half. By world standards, this is robust. (*See* Chapter 7.)

It is fairly difficult to design a fiscal system that actually encourages oil companies to spend more than they would otherwise spend. This would be the ultimate in inefficiency. It is called gold-plating and is a huge concern of many government personnel. The subject always comes up, so there was quite a stir at the end of one presentation when an oil company executive said: "I am really not impressed with PSCs. There is no incentive to be frugal because of 100% cost recovery."

Unfortunately, the speaker was able to escape unpunished because this remark came at the end of an otherwise excellent speech and on the threshold of a long-awaited lunch. But, the statement caused quite a stir and much subsequent discussion. To claim that PSCs do not provide incentives to be frugal is simply wrong, but similar misperceptions continue to circulate.

BOOKING BARRELS

This subject generated lots of discussion at the conference. Laws governing the booking of barrels differ widely among oil-producing nations, adding to the dilemma of deciding the proper course of action. The following comments illustrate the confusion this subject causes, both between oil companies and governments and between rival oil companies.

- Why do some companies (such as Unocal) book *working interest* barrels, while most book *entitlement* barrels under PSCs? Unocal is going to change the way they do it.

- Some companies will *book* gas used as fuel in operations that can amount to upwards of 2% or more. Others claim they do not.

- Who is booking reserves under the *buybacks* in Iran? How can they do that? It looks like TotalFinaElf is booking barrels but indirectly under Middle Eastern operations.

- Most Western companies are legally bound to report (book) reserves, yet some countries do not want any foreign companies *booking barrels* for operations in their country. This has become a big concern.

- Some companies feel that they can book *royalty* barrels if the government exercises their option to take its royalty *in cash* as opposed to *in kind*.

- In countries where taxes are paid *in lieu*, it appears that most companies *gross-up* their actual entitlement as if they had paid taxes directly. This allows them to book more barrels.

- Furthermore, companies are using conversion factors other than 6 MCF per barrel (6:1) to convert gas to crude oil equivalent (COE) or barrels of oil equivalent (BOE). If a company's portfolio has a gas composition relative to oil with a heating value that corresponds to say 5.7 MCF per barrel rather than 6:1 then it can book 5% more BOE for its gas.

TAXES IN LIEU

In numerous countries, income taxes are imbedded in the government share of profit oil (Philippines, Egypt, Syria, Oman, and Trinidad and Tobago, for example). This is sometimes referred to as taxes in lieu. The PSC will state that "taxes will be paid for and on behalf of the Contractor out of the NOC's share of profit oil." This can provide stability because if tax rates are increased, the contractor is immune. These are some of the most stable arrangements in the world.

During a quiet coffee break, some government representatives expressed shock that a Canadian oil company had proposed a *tax in lieu of* arrangement and even went so far as to refer to the Canadians as robbers. I was involved in a similar conversation in one Central

Asian republic last year. When I explained that in fact this is rather common and usually provides some advantages for governments, they were stunned. Countries with these taxes in lieu have some of the highest ERRs in the world.

GOVERNMENT DISBURSEMENT OF TAKE

The means by which governments take—it is probably fair to say—is a bit more highly evolved than the means by which governments disburse. Citizens of this world are no longer as passive about seeking their fair share as they once were, and in some locales they have threatened uprisings and worse. Extreme examples of this problem are found nearly everywhere:

- **Colombia:** In Colombia "vacuna de la gurilla" (guerilla vaccine, i.e., payments, help, assistance to local folks and rebels that helps an oil company get along out in the nether regions of a country) is illegal, but how are companies to deal with the locals?

- **Ecuador:** The Uwa tribe has threatened to commit mass suicide.

- **Nigeria:** There has been so much heartbreak caused by the execution of Ken Sarowiwo, among other things. The Ogoni people are angry. They aren't the only ones. The government is concerned with how to disburse funds yet not increase inflation. The locals say: "Just give us the money." The government worries: "Who do we give it to? Will the Chief disburse any of this?"

- **The list goes on:** Papua New Guinea, Indonesia, the former Soviet Union, California, Venezuela, Angola, and Myanmar.

WHICH SYSTEM IS BEST?

For all practical purposes, there are two main systems around the world: R/T systems or production-sharing systems. Debate continues about the virtues of both, yet the similarities are overwhelming from an economic, accounting, and financial point of view, which perhaps means (some of us believe) that the choice of system may not be such a critical issue. Regardless of this, PSCs have become the fiscal system of choice for most countries. Yet it is significant that in 2000 Algeria has turned away from PSCs and has begun using a R/T arrangement. Saudi Arabia is negotiating three huge arrangements that supposedly will be structured along the basis of an R/T system. It will not be that simple; these deals are huge (ranging from upstream to way-downstream) and difficult to negotiate. Numerous precedents may have to be established, always dangerous for big-dollar contracts, and to refer to these as "big-dollar" is one monstrous understatement.

Imagine a contract that could involve the equivalent of, say, 1 billion BBLS of oil or crude oil equivalent (say 3 TCF of gas or so). These are not *large* by Saudi standards—I am just trying to make a point. If we assume an average price of $25/BOE, gross revenues over the life of a project would be on the order of $25 billion. Now, if total full-cycle costs (Capex and Opex) relative to gross revenues are on the order of 12% (or $3/BOE) then total economic profits are $22 billion. Every single percentage point of take (either government or contractor) comes to $220 million. The difference between an Indonesian/Malaysian government take of, say, 85% or so and an Iranian *buy-back* type contract with government take on the order of 93+% would represent roughly $1.76 billion.

Are we being fair? Well, perhaps it is not entirely appropriate to compare unsigned exploration contracts (although I use the term *exploration* here loosely) in Saudi Arabia with the buybacks in Iran, but the point is there is a lot at stake. And there are some amazing similarities between these contracts (and many others). And there are more than a billion barrels involved.

FINGER ON THE PULSE

Gatherings such as the one in Thailand provide an excellent forum for discussing key concerns of both oil companies and governments. Oil companies are frustrated and accuse governments of being too greedy and/or never taking into account *risk*. I believe the accusations of greed are rather pathetic and usually misplaced. To claim that governments are not aware of risk is not fair as a broad generalization. Governments directly accommodate risk in numerous ways: royalties, cost recovery limits, ringfencing, relinquishment, government participation, and the choice of block size. The discussions also indicate that the science of fiscal system analysis and design is evolving but perhaps still lags behind some of the more highly evolved sciences such as petroleum engineering and even geology. Geologists rarely agree, but at least they can communicate with each other. The combination of misperceptions and diverse non-standard terminology, frustrations, and suspicion creates enough confusion to keep a 2½–day conference exciting from start to finish.

12

Kashagan and Tengiz—
Castor and Pollux

The recently announced super giant Kashagan discovery in the Kazakhstan sector of the North Caspian Sea is the world's largest discovery in three decades. Kashagan, located in shallow water, is an analog to the onshore Tengiz field located approximately 130 to 150 km (85 miles) to the southeast.

Kashagan and Tengiz are the two largest fields in Kazakhstan— their oil reserves alone rival the United States' 22 billion barrels of oil, yet they have hardly begun to produce. Tengiz in 10 years of production has produced less than 10% of its recoverable reserves. There are similar structures in the Kashagan license area that are yet undrilled. Overall, the development costs will likely be 10s of billions of dollars, but revenues to the contractor group (the oil companies) and the Kazakhstan government could exceed *one trillion dollars*. The Kashagan prospect (Figure 12-1), named after the great Kazakh poet, was identified by the Soviets in the early 1970s. However, the extremely promising prospect—located in an environmentally sensitive and high-cost environment—was not drilled at that time.

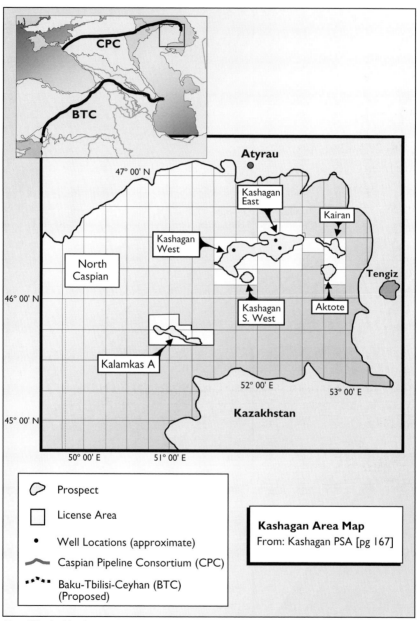

Fig. 12–1 Kashagan area map

Three wells have been drilled on the structure since late 2000, and the prospect has lived up to its promise. Appropriately, the Kashagan PSA is about as famous as the discovery.[1] Every single percentage point of take (either government or contractor) could represent from $1.5 to $2 billion in profits for the first 10 billion barrels alone. The discovery is rated at 6.4 to 100 billion barrels.[2] However, a good working range might be somewhere on the order of 6.4 to 20 billion barrels of recoverable oil reserves—only three wells have been drilled. At 20 billion barrels (if that is ultimately the figure) Kashagan would be the fifth largest oilfield in the world, and the only one of the five outside the Arabian/Persian Gulf region.

The discovery well Kashagan East-1 (KE-1), 47 miles southeast of Atyrau in 10 feet of water, encountered Paleozoic carbonates below 13,000 feet (3960 m) and tested 3700 BOPD and 7 MMCFD on a half-inch choke.[3] In May 2001, ExxonMobil announced test rates on the Kashagan West-1 (KW-1) well located 40–48 km away (depending on the source) from the Kashagan East discovery well. The KW-1 test was also from Paleozoic carbonates (limestone) below 13,800 ft (4250 m) described as a carboniferous-Devonian coral atoll. The well flowed 3300 BOPD of light 42–45° API gravity crude and 7.5 MMCFD on a half-inch choke.[4] October, 2001 AGIP announced a test rate of 7400 BOPD from the Kashagan East-2 well drilled 8 km away from the discovery well.[5] The results of the three wells are summarized in Table 12–1. Also, a detailed chronology is provided in Appendix 12–A.

Table 12–1 Kashagan Drilling Summary

Well	Operator	Well Test Rates		Water Depth (feet)	Drilling Depth (feet)	Comments
		BOPD	MMCF			
KE-1	OKIOC	3,700	7	10	13,000+	Completed Aug. 2000 42-44° AP 1,900 cubic feet per barrel
KW-1	OKIOC	3,400	7.6	22	13,800+	Completed early 2001 25 miles west of KE-1 42-45° AP 2,200 cubic feet per barrel
KE-2	Agip KCO	7,400	-	-	-	Spudded April 2001 5 miles north of KE-1 Results announced Oct., 2001

From: Phillips Petroleum Company *2001 Fact Book*, p. 27; OKIOC website Oct. 23, 2001; Agip-ENI announcement 22 October, 2001 San Donato Milanese

These are not bad test rates, but they are not spectacular by world standards. Of the nearly 100-odd discoveries reported worldwide each year, the average test rate is around 5000 BOPD. The average rate for the upper 25th percentile is 10,000 BOPD, which is the kind of rate that might be expected for a giant discovery like Kashagan. However, indications are that testing has been limited by technical and environmental conditions and regulations.

According to Agip-ENI (the operator), the second well encountered the same reservoir rocks as the Kashagan East-1 discovery well. They estimate deliverability for the well at 5000–20,000 BOPD.[6] This sounds reasonable if the field is as big as expected. This part of the world is famous for *hype*, but all indications are this is a substantial discovery.

TENGIZ

Tengiz is the Kashagan twin. They have the same reservoir rocks with similar fluid properties, pressure gradients, reservoir depths, and sulfur content (*see* Table 12-2). Recoverable reserves are rated at 6-9 billion barrels of light oil (out of 24 billion barrels in-place) with associated gas reserves of 64 TCF.[7]

Table 12–2 Kashagan-Tengiz Vital Statistics

	Kashagan	Tengiz
Discovered	July, 2000	1979
Start-up	N/A	1991
Recoverable reserves (billions of barrels)	6.4 – 20 (or more?)	6 – 9
Location	Offshore 10-22 feet of water	Onshore
Size (acres) (potential productive area)	320,000	100,000
Reservoir depth (feet)	13-14,000	Roughly 14,000 or so (exact figure not available)
Crude characteristics	42-45_ API Gravity 18-20 mol % H_2S	48.2_ API Gravity 12.5 mol % H_2S
Pressure gradient	Assumed to be roughly the same as Tengiz— very high	Very high approximately 0.82 psi/ft +
Gas/oil ratio (cubic feet per barrel)	1,900 - 2,200 from KE-1 & KW-1 tests	High (exact figure not available)
Current (pre-BP/Statoil sale) Working interest ownership (%) (Percentages are rounded) (1) Operator (2) Pre-sale (BP/Statoil) % (3) Joined later	Agip-ENI (1) 14.28% TotalFinaElf 14.28 Exxon/Mobil 14.28 British Gas 14.28 Shell 14.28 BP 9.52 (2) Statoil 4.76 (2) Phillips 7.14 (3) Inpex 7.14 (3)	Chevron (1) 45% Exxon/Mobil 25 Kazakhoil 25 Lukoil 5
Current pipeline ownership (%)	Baku-Tbilisi-Ceyhan (BTC)	Caspian Pipeline Consortium (CPC)
	SOCAR 45% BP (Operator) 25.7- Agip-ENI 5 Other ?	Russian Federation: 24%; Chevron: 15; Kazakhstan Republic: 19; LukArco: 12.5; Shell/Rosneft: 7.5; Exxon/Mobil: 7.5; Sul. of Oman: 7; British Gas: 2: Agip-ENI: 2; ORYX: 1.75; Kazakh Pipeline Vent.: 1.75

Tengiz means *sea* in the Kazakh language, which is a bit ironic because the field is both an onshore field and for all practical (marketing) purposes, it is landlocked. It was discovered in 1979 and began producing in 1991. In April 1993 shortly after the Tengiz contract was signed, the field was producing 24,000 BOPD from 27 Soviet-era wells.

Production costs for Tengiz are reported at around $3/BBL. This is not cheap, especially for a giant oilfield. World average production costs for smaller fields are $3.50 to $4.50/BBL. There is economy-of-scale with the giant Tengiz field, but the harsh technical difficulties neutralize some of that. The Tengiz crude has a specific gravity of 0.787 grams per cubic centimeter– 48.2° API with 0.49% by weight (wt %) sulfur and also has abundant solid bitumen. The associated gas has 12.5 mol % hydrogen sulfide (H_2S).[8]

As recently as 1999, two-thirds of the Tengiz production (around 160,000 out of 250,000 BOPD) went out by rail. The Caspian Pipeline Consortium (CPC) leased 10,000 tank cars, sending up to six trains per day to Russian ports on the Black Sea. This is one of the most expensive means of transportation. Transporting oil to the Black Sea costs around $6/BBL with much of the production going by rail. Transportation cost on the CPC pipeline from Tengiz to the Black Sea port of Novorossisk, which started up in August 2001, is estimated at $3/BBL. The $2.6 billion, 950-mile CPC line has an initial capacity of 560,000 BOPD. Ultimate capacity for this line is 1.5 MMBOPD. Kashagan crude will have to find its own way out.

But there is movement on that front. While ChevronTexaco is negotiating with SOCAR (the Azerbaijan NOC) to purchase a share in the $2.4 billion Baku-Tbilishi-Ceyhan (BTC) pipeline project, Agip-ENI has already purchased a 5% share.[9] This project now seems to be a certainty with strong pressure from the U.S. government and the kind of deliverability expected from Kashagan. It is a question of time, but the pressure is intense now with the Kashagan discovery.

Climate and Infrastructure

This landlocked region is characterized by extreme weather, summer high temperatures on the order of 44° C (110° F) and winter lows of -40° C (-40° F). It is the same latitude as Billings, Montana, but 100 feet below sea level. Ice problems are expected in winter, but year-round-drilling is planned. Infrastructure in this remote part of the world is weak for the world-class development contemplated for Kashagan even with Tengiz nearby. However, with reserves like these, even a large world-class pipeline like the BTC project at $2.4 billion (capital costs) becomes feasible. With 10 billion barrels of oil, this amounts to only around 24¢/BBL.

Reservoir Depths and Pressures

The depth of the reservoir rocks ranges from 13,000 feet to more than 15,000 feet. This is not terribly deep by world standards, but the cost of a 15,000-foot well can easily be twice that of a 10,000-foot well. That extra mile makes a big difference. What makes an even greater difference is the reservoir pressure. Pressures throughout the Caspian region are nearly double that of normal hydrostatic pressure and sometimes more. Tengiz is famous for its high temperature and pressure. Temperatures are nearly 200° F, and pressures are among the highest in the world at 0.82 pounds per square inch per foot (PSI/ft) or more—almost twice normal hydrostatic pressure of 0.433 to 0.465 PSI/ft.

A pressure gradient like this can easily add more than $10 MM per well for drilling fluids (mud) alone, and with the kind of mud weights required (more than 16 pounds per gallon), drilling can go slow. The reported cost for the first two Kashagan wells is US $100 MM, not including the cost of the initial 110,000 square kilometer 3-D seismic data acquisition program that preceded drilling. This does not sound unreasonable. High temperatures and pressures, sour (hydrogen sulfide bearing) gas, high GORs, and poor infrastructure in a hostile and environmentally sensitive region all add up. It will take hundreds of development wells to ascertain the potential of Kashagan, and development wells will not cost as much as exploration wells.

Productive Area

While there are numerous reported sizes for the two fields, a good working number for the Kashagan field is probably 320,000 acres (based on numerous reported figures). There are many exploration blocks/licenses in this world that are smaller. Tengiz is about a third as large. Alaska's Prudhoe Bay field productive area is roughly 150,000 acres. In the environmentally sensitive Caspian, dealing with such a large structure is expected to be a bit of a problem—it takes a lot of wells to cover that kind of area.

COMMERCIAL TERMS

It is likely the Kashagan agreement would have been famous if only for the complexity of the terms. Prior to drilling, the prospects were known to have substantial potential, and it appears that particular care was taken to craft commercial terms that might accommodate all possible outcomes. The contract is unique. It has a half dozen *sliding scales* of various types (in Kazakhstan the term *gliding scale* is sometimes used). I have seen no other contract in the world with this degree of flexibility or complexity, but this does not mean the terms are unfair. In fact, the contract terms are extremely progressive—a *back-end-loaded* system. Government share of profits *and* revenues is extremely low at first. Figure 12–2 shows the ERR is only 2% in the early years.

> This flow diagram shows that in the early accounting periods before payment, the **Contractor Group (Oil Companies)** can obtain up to 98% of the revenues (or production) generated. This example assumes costs far exceed Gross Revenues (tax deductions will yield zero taxable income).

Early Years of Production—Single Accounting Period

Gross Revenues
(net-back sales price—transportation costs netted out)

Oil Company (Contractor) Share	$100.00	Government Share
	Royalty 0%	→ $0.00
	$100.00	
$80.00 ← ($Equivalent of Cost Oil)	**Cost Recovery** 80% Limit	
	$20.00 ($ Equivalent of Profit Oil)	
$18.00 ←	**Profit Oil Split** 90/10%	→ $2.00
($0.00) →	**Tax Rate*** 30%	→ $0.00
$98.00	Division of Gross Revenues	$2.00
	Effective Royalty Rate	2% $2.00/$100.00
98% $98.00/$100.00	Company access to Gross Revenues	

* This analysis is based on the assumption that accumulated costs (deductions) exceed revenues, thus the companies are in a no-tax paying position.

Fig. 12–2 Kashagan PSA example calculation—early years

The heart of this contract is a complex formula for the division of profit oil (defined as gross production less cost oil—*see* Figures 12–3 and 12–4 for these and other terms). Contractor share of profit oil is a function of four separate sliding scales. It is defined as the lower of the contractor profit oil share calculated by either an *R* factor, an IRR factor, or a fairly unique two-dimensional *volume factor*. It seems like a slightly paranoid way of trying to ensure there is no money left on-the-table. (*See* Kashagan PSA Summary in Appendix 12–B).

> **In the long run, the government share of revenues, production, and profit oil will increase. Government Take also increases—for the first 10 billion barrels, the average will likely be around 83%.**

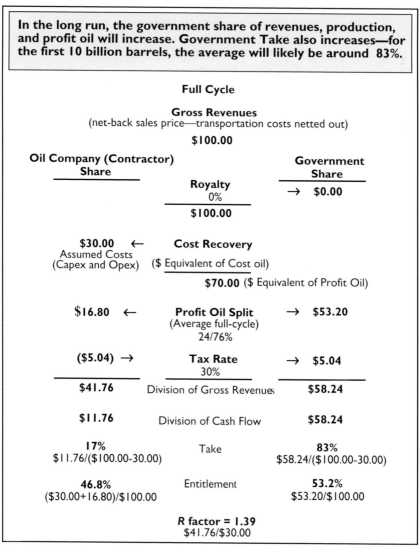

Full Cycle

Gross Revenues
(net-back sales price—transportation costs netted out)
$100.00

Oil Company (Contractor)		**Government**
Share		**Share**
	Royalty	→ $0.00
	0%	
	$100.00	

$30.00 ← **Cost Recovery**
Assumed Costs
(Capex and Opex) ($ Equivalent of Cost oil)

$70.00 ($ Equivalent of Profit Oil)

$16.80 ← **Profit Oil Split** → **$53.20**
(Average full-cycle)
24/76%

($5.04) → **Tax Rate** → $5.04
30%

$41.76   Division of Gross Revenues $58.24

$11.76   Division of Cash Flow $58.24

17%   Take   83%
$11.76/($100.00-30.00) $58.24/($100.00-30.00)

46.8%   Entitlement   53.2%
($30.00+16.80)/$100.00 $53.20/$100.00

R factor = 1.39
$41.76/$30.00

Fig. 12–3 Kashagan PSA example calculation—full cycle

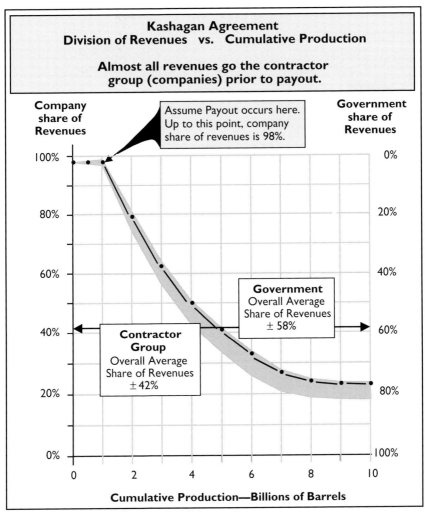

Fig. 12–4 Kashagan Agreement division of revenues

R Factor

The R factor is fairly typical of other such elements around the world with one exception. The R factor is equal to the inflation-adjusted *deflated value* of the contractors' cumulative receipts (effectively contractor cost oil plus profit oil less taxes) divided by the cumulative deflated value of contractor expenditures. It is basically a *payout* formula that adjusts for inflation. The inflation adjustment is unusual. The contractor can receive 90% of the profit oil until the *R* factor reaches 1.4 (inflation-adjusted payout plus 40%). From that point until the contractor reaches an *R* factor of 2.6 (which is unlikely), the contractor profit oil share will slide downward to 10% (unless it has already slid to that point because of the other sliding scales).

The cost recovery limit is also a function of payout—or *payback* as it is referred to in the agreement. It changes from 80% before payback to 55% after—effectively another sliding scale. An *R* factor of one (1.0) is the point at which payout occurs. These formulas are fairly common around the world and are becoming more common. Approximately 25% of the countries around the world have either an *R* factor or an ROR-based system. They rarely have *both* an *R* factor and an ROR formula.

IRR

Systems with taxes or profit oil splits based on various IRR thresholds are referred to as *ROR systems*. They are also referred to as the *World Bank model*. In these systems, pre-determined IRR thresholds are established (by statute or negotiation), and when the contractor's IRR exceeds these thresholds, government take increases by virtue of either an increased tax rate or a change in profit oil split. There are three such sliding scales in this contract. The profit oil, profits tax, and the *volume floor* are all governed by IRR-based sliding scales.

The contractor receives 90% of the profit oil until an IRR of 17% is reached on the contractor's *deflated* net cash flow. The 17% IRR trigger point then represents a *real* ROR—which is common for ROR mechanisms. From that point until the contractor reaches a real IRR of 20%, the contractor profit oil share slides downward to 10% (unless it has already moved to that point because of the *R* factor or the volume factor).

Volume Factor

The third dimension to the profit oil split calculation is a slight variation on the *cumulative production sliding scale* theme. Up to roughly 3 billion barrels of contractors' cumulative share of production (or more precisely a *notional volume* of 3 billion barrels), the contractor share of profit oil is 90%. After that, it slides down to 10% at around 5.5 billion barrels. This aspect of the profit oil split calculation is further qualified by a volume floor limitation. The volume factor profit oil split calculation will be equal to the greater of the notional volume calculation or the volume floor calculation, which is also based on IRR-based thresholds (*See* Kashagan PSA Summary, Appendix 12–B).

The multi-dimensional hybrid nature of the contract is extremely rare. Various simultaneous calculations each yield a profit oil share for the contractor, and one is chosen. The profit oil split calculation applies separately to each development area. The contractor share of profit oil (P/O) at any given time is based on the results of the previous accounting period. An example is shown in Table 12–3.

Table 12–3 Example Contractor P/O Share Calculation

Method	(1) P/O Share	(2) P/O Share	(3) P/O Share
"R factor"	79%	70%	
IRR	58%	58%	58%
Notional volume	55%		
Volume floor	60%	60%	
Ultimate contractor P/O share			58%

(1) Each of four calculations of contractor share of profit oil (P/O) are made.
(2) The choice between either the notional volume or the volume floor calculation will be based upon whichever is "greater".
(3) The choice between the three remaining percentages will be based upon whichever is "lower". In the following accounting period the contractor share of profit oil will be 58%.

WHAT DOES IT ALL MEAN?

The contract terms provide an extremely liberal means by which those who put up the capital can get their money back quickly.

Furthermore, in the early stages, the contractor gets a healthy share of profits. Only later, after costs have been recovered *and* the contractor group reaps a reasonable ROR, does the government take kick into high gear.

Recovering Costs

One of the key aspects of this system is that those providing the capital can recover costs quickly. In any given accounting period, the contractor group can receive up to 98% of the production. The cost recovery limit is 80%, which forces 20% of the production into the profit oil split, and the companies get 90% of that. This means the share of production for the contractor group is 98% [80% + (90% of 20%)]. There is a big difference for companies able to access 98% of production as opposed to only 80% (world average)—especially with so much capital involved.

Government Take

These numerous simultaneous calculations alone make the Kashagan contract unique, and while much of the rest of the contract is based on standard formulas found in other contracts around the world, it is the *percentages* that provide added flavor. The general mechanics of the system are shown in Figure 12–3 where $100 of gross revenues are used to show the general distribution of revenues and profits. It is assumed that capital and operating costs (full-cycle) are assumed to amount to 30% of total revenues for 10 billion barrels of production. It is also assumed that the overall average government share of profit oil during this time will be around 76% (contractor share is 24%). When the tax rate is factored in, the government take comes to around 83%.

Figure 12–5 illustrates how government take changes with time as measured by cumulative production for both Kashagan and Tengiz. The Crazy Horse deepwater discovery in the U.S. Gulf of Mexico is also shown on this graph for counterpoint. Notice the government take at Kashagan is quite low at first and climbs to around 94%.

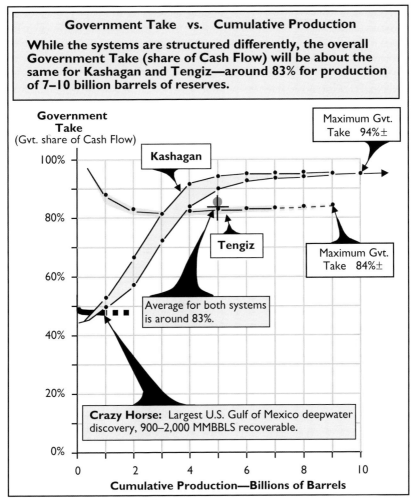

Fig. 12–5 Kashagan and Tengiz agreements

The structures of the Kashagan and Tengiz systems are dramatically different, but for the first 10 billion barrels or so, overall government take is about the same—around 83%. Yet government take at Kashagan is quite low at first—around 46% and increases after the companies reach payout (*see* Figure 12–5). These may sound like tough terms, but by contrast, government take in the famous Indonesian standard contract for oil is around 86–87%. Many companies made a lot of money under those contracts and so did the Indonesian government.

Contractor Entitlement—Booking Barrels

In the early stages of production, the contractor group will be entitled to *lift* a huge percentage of production—98%. In this type of contract, the lifting entitlement will correspond to the reserves the partners will be able to book. Thus for the first 3 billion barrels of (proved) reserves, it is likely that each company will be able to book upwards of 80 to 90% of their working interest share of proved reserves under U.S. SEC guidelines. Under London Stock Exchange guidelines, British companies will be able to book the same percentage of both *proved* plus *probable* reserves ("P_{50} reserves"—considered to have a 50% chance of being too high and/or a 50% chance of being too low). In the example in Figure 12–4, the contractor group entitlement (cost oil + profit oil) comes to 47%. As government share of profit oil increases, so does government entitlement. The government receives profit oil in *kind* and taxes in *cash*.

WHAT IS IT WORTH?

The petroleum industry has some of the greatest contrast between risk and reward. When discoveries are made, the curiosity about *value* becomes magnified. Without detailed information, the transactions are difficult to evaluate fully, but it is always interesting to try. Actual market transactions, if they are arms-length and conducted in a competitive and efficient environment, can be the acid test of value.

In 1998 Phillips and Indonesian Petroleum Company (Inpex) acquired the Kasakh Government 14.28% share (7.14% each) for $500 million cash plus other consideration including low-interest-rate loans and so forth.[10] This transaction indicates a value at that time of $3.5 + billion, but this was before any drilling had taken place.

In early 2001 BP agreed to sell its 9.52% interest in Kashagan to TotalFinaElf (TFE) for $409 MM. This agreement took place after the results of the first well (Kashagan East-1), but before the results of the second. At about this same time, Statoil also agreed to sell their 4.76% share to TFE for $221 MM. (*See* Kashagan Chronology, Appendix 12–A). These proposed transactions indicate a value for the discovery (and ancillary interests) of $4.4 billion.

Proposed Sale of BP/Statoil Interests:

	Working Interest Share	Sale Price US$MM	$MM per 1%
BP	9.52%	$409	43
Statoil	4.76%	$221	46
Total	14.28%	$630	44

Assuming there are 10 billion barrels of recoverable reserves, the value comes to around 44¢/BBL for partially *proved* and *probable* undeveloped barrels in the ground. However, the proposed deal is fraught with controversy. There are many who believe this proposed transaction does not represent a valid indication of actual value. In fact, shareholders have filed a class-action suit against the BP Board of Directors claiming "gross waste of assets" in the sale of BP's 9.52% interest in Kashagan.[11]

Furthermore, this deal was not consummated because the other partners had rights of first refusal. By August 2001, all participants indicated their intention to preempt the sale and take their respective shares at this price.[12] None of them considered the price too high. The total consideration in this proposed transaction is similar to that of the earlier (pre-discovery) Phillips/Inpex acquisition, but there are dramatic differences.

Considering the change in oil prices and the fact that the Phillips/Inpex transaction occurred *before* much of the hydrocarbon potential of the Kashagan prospect had been confirmed, the price for the proposed BP and Statoil transactions seems low. A comparison is provided in Table 12–4.

Table 12–4 Comparison of Kashagan Transactions

Buyer	Phillips/Inpex	TotalFinaElf* (originally intended)
Seller	JSC KazakhstanCaspiShelf	BP and Statoil (separate "proposed" transactions)
Acquired working interest	14.28%	9.52% and 4.76%
Price paid (or proposed)	$500 million+	$630 million
Imputed value of 100% working interest	> $3.5 billion	$4.4 billion
Historical setting		
Transaction date	1998	2001**
Oil price at the time	$11/bbl	$20+/bbl
Other	*Before* the discovery was made	*After* the discovery was made and appraised with a second well

* Ultimately all other partners exercised right of first refusal August 2001
** Announced in February, finalized in June 2001, preempted in August—not finalized

Under a variety of scenarios, the value of the interests in the Kashagan license could be worth many times the price paid by Phillips/Inpex. This often happens when a big discovery is made. It all depends on oil prices, costs, and timing—an age-old formula. With present value discounting, depending on the cost, price, and timing assumptions, the value per barrel at this point is something on the order of $1/BBL. This would put a value on the discovery, (assuming 10 billion barrels) of around $10 billion.

In addition to this, there are other prospects in this acreage: Kalamkas, Aktote, Kairan, and Kashagan Southwest. Any one of these could be another world-class discovery (*see* Figure 12–1 Kashagan Area Map from the Kashagan Agreement). It will be fascinating to see the drilling results. Furthermore, there is a substantial amount of gas. With a GOR of around 2000 cu ft per barrel (based on test rates), there could be at least 20 trillion cu ft of *associated gas* alone (based on 10 billion barrels recoverable reserves). This would add another 3.3 billion BOE, but it is extremely likely that the gas reserves will be substantially greater than this. The area is exciting from any perspective. Kashagan will be big news for many years to come.

On October 10, 2002 TotalFinaElf issued a press release announcing that the first Kalamkas-1 well, had successfully been tested. The well was drilled to a total depth of 2360 meters and encountered several hydrocarbon reservoirs in the Jurassic section. The well tested an initial flow rate of 2300 BOPD.

REFERENCES

[1] Gordon Barrows, The Barrow Company, New York.

[2] Approximately 10–60 billion barrels (Reuters, October 9, 2000); 50 billion barrels (*International Herald Tribune*, November 15, 2000); 13–14 billion tonnes of oil, approximately 100 billion barrels (*ITAR/TASS* News, February 2, 2001; 6.4–10 billion barrels, "New world oil reserves..." (*IHS Energy Report*, Reuters, July 12, 2001); 30 billion barrels (*The Guardian*, February 3, 2001); 25–60 billion barrels (*Oil & Gas Journal*, July 16, 2001, page 37).

[3] Phillips Petroleum Company *2001 Fact Book*, p. 27.

[4] OKIOC website, October 23, 2001.

[5] Agip-ENI announcement, October 22, 2001

[6] *Offshore Magazine*, August 2001, p. 136.

[7] Connell, D., *Oil & Gas Journal*, vol. 98, issue, 8, p. 136.

[8] Ibid.

[9] "Oil giant holds talks over BTC pipeline," *The Russia Journal*, November 2–8, 2001.

[10] LaVine, S., "Phillips Petroleum and Japanese in Big Kazakh Exploration Deal," *The New York Times* website, September 15, 1998.

[11] *Platts Oilgram News*, July 23, 2001.

[12] Philips Petroleum Company *2001 Fact Book*, p. 28.

13

The Bidding Dilemma:
a 20-Year Retrospective

International petroleum exploration is extremely challenging these days. While the amount of exploration acreage available worldwide has more than tripled in the past 15 years, there are also more companies seeking opportunities than ever before. From the point of view of most governments, this is a healthy environment. However, it has not been healthy for most companies. For the past two decades, the exploration end of the business has been notoriously unprofitable. Part of the reason is that fiscal terms are so onerous in most countries. This is because the industry has been plagued by chronic overbidding that has shaped the market for exploration acreage and projects. Bidding and/or negotiations in the industry have been strongly influenced by both increased competition and overoptimistic estimates of oil prices, costs, prospect sizes, and success ratios.

POSTMORTEM ANALYSIS

Postmortem analysis of exploration portfolios of the 1980s and 1990s shows consistently overoptimistic estimates of two key variables: prospect sizes and success ratios. For example, for any portfolio of prospects, there should be both an average prospect size and an associated average probability of success. Assume we are performing an analysis of the results of 10-years of exploration during which 100 prospects were drilled with an average size of 90 MMBBLs and an estimated probability of success of, on average, 20%. If we evaluate the actual discoveries that resulted from the 100 prospects drilled, there should have been roughly 20 discoveries and 1.8 billion barrels of reserves discovered (an average size of 90 MMBBLS each). However, typically and consistently, the results of a postmortem analysis like this for the 1980s and 1990s are substantially less exciting than expectations. Actual success ratios are lower, and the average discovery is smaller than expected.

Unfortunately these overexpectations provided the basis for numerous bids and negotiations during the last two decades. This cannot help but result in overbidding and, ultimately, loss of value:

All in all, such exploration for new giant fields destroyed value rather than creating it in the 1980s and early 1990s… Exploration, as a corporate function, lost credibility. (Rose, 1999).

A McKinsey & Company report estimated for the petroleum industry:

$400 billion value destruction over the 1980s (Conn and White, 1994).

Most of this loss was from exploration. Similar conclusions exist for the bonus bidding in the U.S. federal offshore.

In 1970 after about a decade and a half playing this gambling game, the estimate was that bidders were more than $4 billion in "deficit". After about three decades, our estimate is that bidders are about $48 billion behind. (Lohrenz and Dougherty 1983).

This $48 billion estimated deficit is interesting, considering cumulative bonuses for the U.S. outer continental shelf (OCS) were only $47 billion by 1983. The statement by Lohrenz and Dougherty would imply that any bonus was an overbid. And this conclusion is shared by this headline:

U. S. OCS Operators in the Hole by $70–80 billion (Warren, 1989).

Cumulative bonuses for the U.S. OCS were only $55 billion in 1989 so even a *zero* bid would presumably have been too high.

Additional analysis of the Gulf of Mexico, studies done through 1982 before the U.S. Minerals Management Service went *area-wide* indicated that:

The average block had three bidders and the average winning bid was $8 million. The second highest bid averaged $3.2 million and the lowest bid was $1.4 million. (Warren, 1989).

It is extremely common in bonus bidding situations that the highest bid is about two times greater than the next highest bid. Almost all the literature dealing with bidding performance such as that outlined above will refer to the highest bid as an overbid. Typically too, the amount of overbidding quoted in the literature will correspond to the difference between the highest bid and the next highest bid (the money left on the table). In my opinion, it is more likely that in situations where there are multiple bids, the highest bid is not the only overbid.

A classic example would be a single bid of $5 MM. This is considered an overbid of $5 MM relative to zero. Why zero? The *real value*, considering prospectivity, oil prices, and fiscal terms could have been less than zero. If royalties and taxes are too high relative to the geopotential of a province, then even *no bonus* (zero) is too high.

OIL PRICE ESTIMATES

Our industry has been dramatically overestimating oil prices for the past 20 years. This alone provides a substantial bias. Figure 13–1 is a common type of illustration of actual versus expected prices over the years. Expectations have not matched reality. Unfortunately *expectations* are what drive competitive bidding—not *reality*.

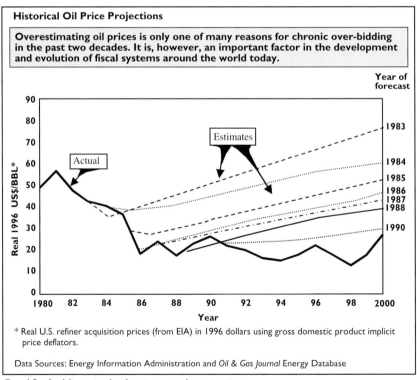

Fig. 13–1 Historical oil prices and projections

Cost and Timing Estimates

In my experience, overoptimism in estimating costs may not be as consistent a problem as others, but it is a problem. Going over-budget seems to happen a lot more often than bringing projects in on time and under budget—particularly with mega-projects and frontier regions.

The industry has dramatically reduced time requirements to get from discovery to startup and then to peak production. Still, with hindsight, we find our estimates to mobilize large-scale exploration efforts and/or development projects into remote provinces were usually overly optimistic.

Prospect Sizes

Overestimating reserve potential of an un-drilled structure (a *prospect*) is an extremely common problem in the industry.

> ...*it must be acknowledged that overestimation of prospect reserves is a widespread industry bias that has proved difficult to eliminate* [Johns et al., 1998; Alexander and Lohr, 1998; Harper, 1999] (Rose 2001).

Notice the dates associated with the previous statement. The industry became acutely aware of the problem in the late 1980s and early 1990s. While many company personnel these days feel that much of the problem has been resolved, it is still a problem. One of the main cures is internal peer review and *team* approaches to exploration. Table 13–1 summarizes the results of studies that evaluated the difference between estimated versus actual reserves. The overestimates ranged from 30% to more than 160%. One of the most common explanations for this tendency is that geologists and explorationists must be optimistic in order to sell their projects and compete for funds internally.

Table 13–1   Examples of Reserves Prediction Accuracy*

	Overestimated by:
Various, Rose, 2001	30–80% (1)
Gulf of Mexico, Capen, 1992	100% (roughly)
Norway, 8–14th rounds, Rose, 2001	163% (2)
Deepwater, BP-Amoco, 15-year retrospective since early 1990s Harper, 1999	122%

(1) "Since 1993, most oil companies have acknowledged that their geotechnical staffs persistently overestimate prospect reserves, commonly by about 30% to 80%" (Rose, 2001).

(2) For example, if a company estimated a prospect at 200 MMBBLs but the discovery yields 76 MMBBLs, the estimated exceeded reality by 163% [(200-76)/76].

* Summarized from: Peter R. Rose, "Risk Analysis and Management of Petroleum Exploration Ventures," AAPG *Methods in Exploration Series*, Nov. 12, 2001.

SUCCESS RATIOS

Estimating the probability of success (some refer to this as *chance factor*) is an absolutely critical element in exploration risk analysis. And most companies have been making these estimates directly (EV analysis) or indirectly (gut feel) for decades. Either way, we have been overestimating, and it is killing us.

Almost pro forma, explorers use 10% to 15% for high-risk prospects; in reality, however, most should be 1% to 5%. (Forrest, 2002).

One of the larger oil companies in the mid-1990s decided to implement a new strategy because of the failures of the 1980s and early 1990s. The new exploration strategy was based on a decision to avoid further *high-risk* exploration. No more exploration would be undertaken unless the probability of success was greater than 20%. The hurdle rate (target ROR) was set at 15%. Over the next five years, overall exploration success increased to around 45%, but it was acknowledged internally that the 15% target IRR threshold was not being met. In fact, the investments were not even obtaining an IRR equal to corporate cost of capital. In other words, value was not being added. If value is not being added, then it is being eroded. The five years of exploration represented hundreds of millions of dollars of investment. There are numerous stories like this.

Four Examples—Perspectives on Competitive Bidding

Four independent examples are provided to illustrate key aspects of the development and/or analysis of competitive bidding/negotiations in petroleum exploration. There is almost always a determining factor of some sort including bonus bidding, work program bidding, fiscal terms such as royalties or profit oil spits, or combinations of these elements. The variety of means by which governments allocate licenses is diverse. Examples #1 and #2 illustrate examples of where the art and the science of bidding are put to the test. Example #1 gives an example of pure bonus bidding and how bids are developed in these situations. Example #2 illustrates a situation where *terms* are the bid item.

Example #3 shows a combination of both bonus and terms as bid items. Ranking combination bids like these require both cash flow analysis and risk analysis. Example #4 is a variation on the basic themes developed in Examples 1–3. It provides yet another perspective on overbidding.

Example #1: Signature Bonus Bidding— Part Art, Part Science

Two workhorses of the exploration business are discounted cash flow (DCF) analysis and EV risk analysis. But these tools, as widely accepted as they are, only provide boundary conditions for a typical competitive bidding situation. Figure 13–2 in Example #1 depicts a simple "two-outcome" model of a possible drilling prospect subject to a competitive bonus bid. For the sake of convenience, simple two-outcome models are used in these examples instead of the more common multi-outcome decision tree models used by almost all companies. The principles though are the same.

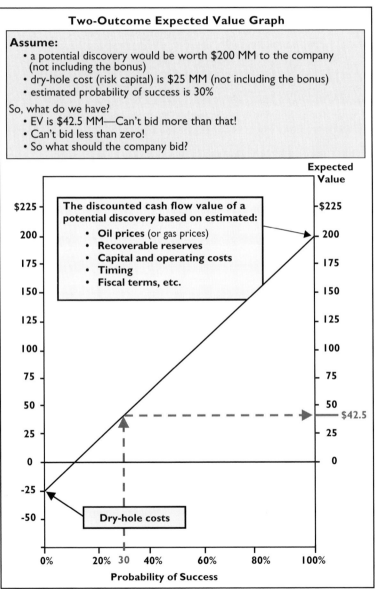

Fig. 13–2 Two-outcome EV graph

In this example, the potential "reward" is based on DCF analysis of the contractor cash flow from a potential discovery which yields an un-risked value of $200 MM (excluding the effect of any bonus). Dry-hole costs (not including a bonus) are estimated at $25 MM. The EV assuming a 30% chance of success is $42.5 MM [($200 MM * 0.3) + (-$25 MM * 0.7)]. The analysis then provides the boundary conditions for a possible bonus bid. The company should not bid more than $42.5 MM nor can it bid less than $0 (zero). So what should it do?

The most credible recommendations suggest bidding some percentage of EV, say, 20–30% or less depending upon the circumstances and expected competition [Capen, et al., 1991 and Rose, 2001]. This would yield a bid of $8–12 MM or less.

The moral here is that it takes both science and art. Defining the boundary conditions is pure by-the-numbers DCF and EV *science*. Deciding what percentage of EV (say 25% to 35% or so) requires a little more *art*. Example #2 is similar in this respect. In provinces such as the Gulf of Mexico, where there is substantial publicly available data, the determination of how much of EV to bid becomes more scientific. But frontier areas are different—less history and less public data.

Note: In many countries, bonuses are tax deductible (but not cost recoverable). This has been ignored here for illustration and discussion purposes.

Example #2: Bidding Terms

Imagine evaluating a block/prospect in order to develop a bid based solely on profit oil split. Assume economic and risk analysis of the prospects on the block with varying levels of profit oil split yield the following relationship between overall government take (which includes the effect of profit oil split, royalties and taxes) and company EV.

Government take (%)	Company Expected Value * ($MM)	
50%	$170 MM	
55%	150 MM	
60%	125 MM	
65%	97 MM	Range for last 10 contracts
67%	85 MM	Average of past 10 contracts
70%	65 MM	
75%	48 MM	
80%	21 MM	
83%	0 MM	Breakeven
85%	-13 MM	
90%	-29 MM	

* Assume the discount rate corresponds to corporate investment criteria *hurdle rate* close to or slightly greater than cost of capital.

What to bid?

1) What if the past 10 contracts in the country had 60–70% government take?

2) Assume the average of the past 10 contracts was 67% government take.

3) What if the geopotential/prospectivity of the last 10 blocks awarded was better than the block you are looking at?

4) Assume oil prices were higher when the last 10 contracts were signed.

An example like this certainly requires *art*. I would not be comfortable with anything greater than 65% government take but what if company management insists on it?

Example #3: Combination Bid—Bonus and Terms

Assume the government will award exploration rights for a particular block to the highest bidder, and both *terms* (profit oil split and/or royalty) and a signature bonus are biddable. The government receives two bids:

Summary of Bids

Company A   Bid #1 $10 MM + Government take of 66%

Company B   Bid #2 $5 MM + Government take of 78%

Bid #1 relative to Bid #2 has a larger bonus but a lower government take. This provides a classic trade-off—part of the government take is guaranteed (the bonus) and part is *at-risk* or at least *uncertain* (it depends upon whether or not a discovery is made). Analysis of these two bids requires DCF analysis and risk analysis. Depending on the prospectivity, either bid may be superior to the other. If the prospectivity is extremely poor, then it is likely the government would prefer Bid #1. On the other hand, if prospectivity is quite good, then the government will likely prefer Bid #2. It depends upon the probability of success and the potential size and/or value of a potential discovery (or discoveries).

Analysis of the two bids is shown in Example #3. It assumes that the chance of success is 30% and that the present value of a discovery (to the government due to royalties, taxes, profit oil etc, based on DCF analysis) would be $250 MM for a government take of 66% and $290 MM for a government take of 78%. The EV for the government is greatest with Bid #2. Example #3 as illustrated in Figure 13–3 shows a different representation of a two-outcome model than the graph in Example #2 (Figure 13–2), but it is the same concept. Notice there is only about an 8% difference between the two bids from this perspective, even though the highest bonus was twice as high as the next. This approach is explored further in Example #4.

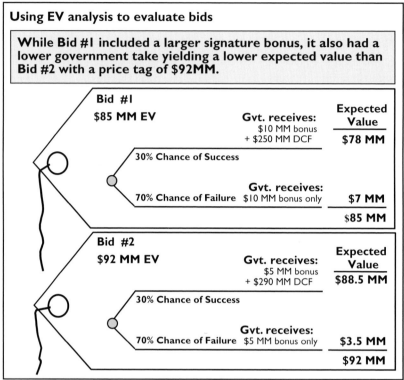

Using EV analysis to evaluate bids

While Bid #1 included a larger signature bonus, it also had a lower government take yielding a lower expected value than Bid #2 with a price tag of $92MM.

Bid #1

$85 MM EV

Gvt. receives:
$10 MM bonus
+ $250 MM DCF

Expected Value

$78 MM

30% Chance of Success

Gvt. receives:
70% Chance of Failure $10 MM bonus only

$7 MM

$85 MM

Bid #2

$92 MM EV

Gvt. receives:
$5 MM bonus
+ $290 MM DCF

Expected Value

$88.5 MM

30% Chance of Success

Gvt. receives:
70% Chance of Failure $5 MM bonus only

$3.5 MM

$92 MM

Fig. 13–3 Using DCF and EV analysis to evaluate bids

Example #4: Bonus Bids—Another Viewpoint

In this example, the government will award exploration rights for a particular block to the company submitting the highest bonus bid. Other terms such as royalties and taxes are *fixed*. The fixed terms yield a government take of roughly 66% (*see* Table 13–2). The government receives three bids:

Summary of Bids

Company A $10 MM

Company B $5 MM

Company C $3 MM

Table 13–2 Example #4: Summary of Assumptions—Economic Model

Field size	100 MMBBLs
Peak production rate	30,000 BOPD
Oil price	$18.00 (flat—no escalation)
Capex	$3.50/BBL
Opex	$3.00/BBL
Government take	66% (20% royalty 50/50% profit oil split) $760 MM (undiscounted) $250 MM (discounted at 12.5%)
Contractor take	34% $390 MM (undiscounted) $50 MM (discounted at 12.5%)
Probability of success	30%
Dry hole costs	$5 MM (does not include potential bonus)
Company expected value (without signature bonus)	$11.5 MM (does not include potential bonus) [0.3 x $50 MM + 0.7 x ($5)]

Government Expected Value Analysis of the Three Bids

	Bonus Bid	**Govt. EV**	**[Formulas - $MM]**
Company A	$1MM	$85 MM	[(($250 + 10) x 0.3) + ($10 x 0.7)]
Company B	$5MM	$80MM	[(($250 + 5) x 0.3) + ($5 x 0.7)]
Company C	$3MM	$78MM	[(($250 + 3) x 0.3) + ($3 x 0.7)]

Industry literature is rich with terminology and analysis for situations like this. The money left on the table (by Company A) is $5 MM—the difference between the two highest bids.

Some quantify *winner's curse* at $5 MM—same as the money left on the table. Others calculate winner's curse at $4 MM because this was the amount by which the company *overbid* relative to the *average* bid ($6 MM). The reasoning is that theoretically each bidder is competent and has access to roughly the same information as the others. Therefore the average bid could more accurately reflect the true *value* of the block, but fortunately for the government it does not have to take the average bid. And furthermore, the average bid does not necessarily represent the average *value* perceived by the universe of bidders.

What about Companies D and E? Let's say there were two other companies who evaluated the deal but opted not to bid—Companies D and E. Just because they did not bid does not mean they think the property is worth zero ($0). They probably did not submit bids because they believed the value was less than zero. Their opinion unfortunately does not get captured in the bid statistics. Assume Company D and E estimated the (bid) value at -$9 and -$13 MM respectively. In other words, Company D would not be willing to take on the block unless the government paid it $9 MM. "The only way we would consider the deal is if the government paid us to drill it!" That of course is crazy, so that is why no bid is submitted. Analysis of all the various opinions of the value of the block then looks like this:

Perceived Value

Company A	$ 10 MM Bid
Company B	$ 5 MM Bid
Company C	$ 3 MM Bid
Company D	$ -9 MM No bid
Company E	$ -14 MM No bid
Average	$ -1 MM

All the bids are too high relative to the average perceived value.

EV analysis of the bids from the government point of view indicates a difference in value of only about 6% between the highest and the lowest bids. This is based on the assumption that there is a 100 MMBBL potential and a 30% chance of achieving that potential (*see* Table 13–2). The EV of the highest bid—$85 MM—only exceeds the next highest bid by $5 MM.

If the highest bid is too high (by only 6%), then perhaps the next bid is also too high. If that is true—and industry performance appears to indicate that it is—then the money left on the table perhaps exceeds $5 MM. Winner's curse is greater than $5 MM. A zero ($0) bonus bid would likely represent an overbid—by at least $4 MM.

CONCLUSIONS

The development and evolution of fiscal terms worldwide in the past 20+ years has taken place in an environment of intense competition and overoptimistic expectations. It has resulted in terms that, generally speaking, are simply too tough.

The average government take worldwide is around 67%, but this is too high for average geological potential (or prospectivity). For countries with better-than-average potential, the government take is closer to 80%. However, better-than-average geological potential is rarely sufficient to sustain such a high government take.

Certainly many countries have modified and/or improved their terms, but relative to the dwindling prospectivity as geological basins have matured in the past two decades, the terms are tougher. This is not because greedy governments have forced these terms on an unwilling industry. It is the other way around. Industry has determined what the market can bear, and it is almost unbearable. Governments have little choice than to allow a competitive marketplace to work its magic as they have for many years.

What Is a Company to Do?

Companies appear to have improved the accuracy of their prospect size and success ratio estimates. However, there is still room for improvement. Internal peer review, or even better, third-party reviews can be extremely helpful. Many companies would do well to avoid those countries where allocation strategies magnify the already hyper-competitive nature of the marketplace. When Venezuela offered 10 blocks in January 1996, they allocated each block separately. This added to an already intensely competitive atmosphere.

Also, companies should carefully target their bidding efficiency and be prepared to lose more bids than they have in the past. I would be uncomfortable with a bidding efficiency greater than 20%, i.e., where more than 20% of the bids submitted are the successful (highest) bid.

REFERENCES

Alexander, J. A., and J. R. Lohr, "Risk Analysis: Lessons Learned," (SPE 49030), Presented at Society of Petroleum Engineers Annual Technical Conference and Exhibition, New Orleans, Louisiana, 27–30 September, 1998.

Capen, E. C., R. V. Clapp, and W. M. Campbell, *Competitive Bidding in High-Risk Situations* (SPE 2993), 1971, pp. 206–218.

Capen, E. C., "Dealing With Exploration Uncertainties," Steinmetz, R., (ed), *The Business of Petroleum Exploration: AAPG Treatise of Petroleum Geology—Handbook of Petroleum Geology*, Chapter 5, 1992, pp. 29–61.

Clapp, R. V., and Stibolt, "Useful Measures of Exploration Performance," *Journal of Petroleum Technology*, October 1991, pp. 1252–1257.

Conn, C., and D. White, *Revolution in Upstream Oil and Gas—Strategies for Growth beyond 2000,* McKinsey & Company, Australia, 1994.

Forrest, M., "Phrases Can Raise Decision Results," *AAPG Explorer,* Feb. 2002, p. 31.

Haper, F. G., "BP Prediction Accuracy in Prospect Assessment: A 15-Year Retrospective," reprint of AAPG International Conference paper, Birmingham, England, 1999.

Johnston, D., and D. Johnston, *Economic Modeling Handbook,* University of Dundee, Centre for Energy, Petroleum and Mineral Law and Policy, Dundee, Scotland, 2002.

Jones, D. R., S. G. Squire, and M. J. Ryans, "Measuring Exploration Performance and Improving Exploration from Predictions—with Examples from Santos' Exploration Program," Proceedings of 1998 Australian Petroleum Producing and Exploration Association, Canberra, Australia.

Lohrenz, J., and Dougherty, E. L., "Bonus Bidding and Bottom Lines: Federal Offshore Oil and Gas," SPE Annual Technical Conference and Exhibition, San Francisco, California, October 1983.

Lohrenz, J., "Profitabilities on Federal Offshore Oil and Gas Leases: A Review," *Journal of Petroleum Technology,* June 1988, pp. 760–764.

Megill, R. E., and R. B. Wightman, "The Ubiquitous Overbid," *Oil & Gas Journal,* 4 July, 1983, pp. 121–127.

Rose, P. R., "Analysis is a Risky Proposition," *AAPG Explorer,* March 1999, p. 14.

Rose, P. R., "Risk Analysis and Management of Petroleum Exploration Ventures," American Association of Petroleum Geologists, Methods in Exploration Series, No. 12, 2001.

Tavares, M. J. D., "Bidding Strategy: Reducing the "Money-Left-on-the-Table" in E&P Licensing Opportunity," (SPE 63059), presented at Society of Petroleum Engineers Annual Technical Conference and Exhibition, Dallas, Texas, October 1–4, 2000.

Warren, J. E., 1989. "U.S. OCS Operators in the Hole by $70–80 Billion," excerpts from Offshore Technology Conference luncheon speech, Offshore, June 1989, p. 26–27.

14

Retrospective, Government Take— Not a Perfect Statistic

The time seems appropriate to reflect on the status of the science of petroleum fiscal system analysis and design and the role of the *Petroleum Accounting and Financial Management Journal*. The *Journal* has published 15 articles or columns of mine on the subject over the past nine years. I find it somewhat horrifying to see how much I repeat myself; nevertheless, a number of new concepts were developed, and the *PAFMJ* deserves credit for helping advance the science. We have made progress over the past decade. Among other things, people are better able to communicate.

The fact that there is so much repetition indicates, in part, the embryonic nature of the science. People have not always been able to communicate comfortably on the subject. Most of the repetition is focused on the concept of government take, and as I write this chapter, it is extremely difficult for me to refrain from defining the term yet again.

The take concept is the focus of criticism now and then, and while I intend to level additional criticism, it won't be the first time. The strengths and weaknesses were first discussed in the Summer 1998 issue of the *PAFMJ* [Vol. 17, No. 2, p. 52 (Chapter 4 in this book)] and again in the Summer 2001 issue [Vol. 20, No. 2, p. 120 (Chapter 11 of this book)] where it was illustrated how the harsh Indonesian government take of more than 85% was partially justified by the combination of block size, relinquishment and ringfencing provisions, and the *savings index* (first published and discussed in the *PAFMJ*, Spring 2000, Vol. 19, No. 1, p. 8, (Chapter 7 here).

There are other weaknesses. One of the biggest complaints is the lack of a standardized terminology, particularly regarding take statistics. This prompted me to include a fairly exhaustive list of similar and competing terms in the *PAFMJ*, first in the summer of 1996 (*see* Chapter 2) and again, with some additions, two years later (*see* Chapter 4).

Some feel that too much attention is placed on this measurement, or they feel that while there is some importance, it is minor. For example:

International petroleum economist Daniel Johnston refers to the IOC's share of the revenue as the IOC's "Access to Gross Revenue," and he refers to the Host Government's share of the revenue as the "ERR."

Daniel Johnston refers to the IOC's share of the profit as the "Contractor Take" (also referred to by others as the "IOC Take" or the "Company Take"). He refers to the Host Government's share of the profits as the "Government Take" (also referred to by others as the "State Take").

While most HGCs are from 50 to 200 pages in length (and the associated petroleum legislation, petroleum regulations, foreign investment laws and other elements of the applicable UPR often are even more voluminous), normally only four to eight pages of an HGC, in addition to pertinent provisions of the tax code, are devoted to the fiscal terms pertaining to the "revenue splits" and "profit splits". The balance of the HGC, and/or the associated pertinent legislation, describes the remainder of the relationship between the Host Government and the IOC. Indeed, there is vastly more at issue in each UPR than the "revenue split" and the "profit split".[1]

(UPR = upstream petroleum regimes, IOC = international oil company, HGC = host government contract)

I do not agree with much of the theme developed in the previous excerpt. If parties to a potential contract (HGC) cannot agree on the appropriate *revenue split* and or *profit split*, then there is no need for crafting or negotiating all that other contract language. There is no deal. As discussed in Chapter 5, there is a balance between prospectivity and fiscal terms that includes among many things the profit split and revenue split. It is not easy finding the proper balance. The exercise is not trivial and involves geophysicists, geologists, petrophysicists, engineers, economists, lawyers, accountants and financial types, and space-age technology.

The attitude reflected in quoted paper reminds me of a project I worked on three years ago. I was working with government representatives from a land-locked country with no oil or gas production. Geological potential was not well understood. Furthermore, no drilling had taken place in nearly 15 years. Costs were expected to be high because all goods and services would have to be brought in from a long way off. There were no indigenous oilfield services of any sort. The country had tendered some blocks in an official license round, but there was virtually no interest—except for one company. This company was coming to the country for an official visit to submit a bid. The government officials and I agreed that a reasonable government take considering the geology and conditions would be around 50% or possibly even less. This would be consistent with frontier terms around the world, and we considered these blocks to be in this category.

To our surprise, the company offered terms that yielded a government take of around 70%. World average is less than that, and these were not world average rocks or conditions—they were worse. The government representatives had no choice but to accept the offer. Their hands were tied. Since then, the company has been unable to find partners—the terms are too tough. Governments know that finding partners is part of the service that companies provide; they are in effect raising capital for and on behalf of the government this way. Unfortunately, this company blundered. Despite having a good contract with around 200 pages of well-crafted contract language, it just had a bad deal. Now both parties suffer.

In defense of economists, accountants, and financial analysts/advisors who deal with petroleum contracts and negotiations, I believe it is fair to say that few if any confine their analysis of a contract to just "four to eight pages." Most of us read the whole contract. Occasionally there is some debate about various contract elements and just what constitutes an economic or financial issue and what the implications are. Of the four elements in the Indonesian PSC previously mentioned, some are purely financial, but others may not be considered to be so *pure*. Yet their combined effect is certainly financial.

There are numerous other examples: a contract may have a dispute resolution clause that provides for binding international arbitration, in English, in a third-party country, under a recognized convention such as the Paris Convention or equivalent, to which the host government is a signatory with Parliamentary ratification. On the other hand, the contract may not have an *arbitration clause*. Is this a financial issue? I believe it is. It makes a difference. There are many other similar issues.

I believe the government take statistic suffers from both *under*-use and *over*use. When people are unaware of the weaknesses (and I believe few are intimate with all the weaknesses associated with the take statistics) then overuse is extremely likely. Here are some of the problems, as I see it, with the government take statistic:

- It does not explain *how* the government *takes*.
- It does not adequately capture the effect of:
  - Signature bonuses
  - Ringfencing provisions
  - Front-end loading
  - Reserve/lifting entitlements and *ownership*
  - Work program provisions
  - Crypto taxes
  - Time value of money
- Its macroeconomic scope is too narrow.
- It is not relevant in some important situations.

It Does Not Explain *How* the Government *Takes*

This aspect has been fairly adequately developed, particularly in the *PAFMJ* [Summer 1998, Vol. 17, No. 2, p. 52 (Chapter 4 in this book), and Summer 2001, Vol. 20, No. 2, p. 120 (Chapter 11 of this book)] where the concept of crypto taxes, ERR, and others were first developed. Companion statistics and other information provide much of this. The ERR is one of the best. Figure 14–1 captures some of the additional context within which the take statistics fit. Take statistics are simply not stand-alone.

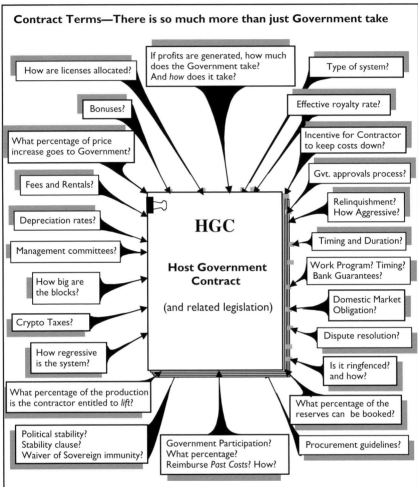

Fig. 14–1 Things to consider

Its Macroeconomic Scope Is Too Narrow

Take does not measure *everything* that matters to a government. The economic impact of the industrial hyperactivity in the U.K. sector of the North Sea, a direct result of the *lenient* terms of the 1990s, is difficult to measure. The *gross benefits* to the U.K.

government go way beyond direct tax revenues and royalties received from the upstream sector of the petroleum industry. Furthermore, the dramatic increase in activity in the U.K. started in the late 1980s and early 1990s when the U.K. government dropped the ringfence for the 75% PRT. This boom occurred *before* the government take (as it is ordinarily measured) was drastically reduced (*see* Table 14–1). The U.K. offshore became the most active offshore province in the world. Reducing the government take in the following years has managed to sustain that boom. Activity and employment in the British petroleum sector is healthy and robust. Government take does not measure the benefits of that.

Table 14–1 U.K. Petroleum Taxation History

Year	Royalty	Supp. Petroleum Duty	Petroleum Revenue Tax	Corporate Tax	Old Fields Marginal Take	New Fields Marginal Take
1974	12.5%			52%	58.0%	58.0%
1975	12.5		45%	52	76.9	76.9
1976	12.5		45	52	76.9	76.9
1977	12.5		45	52	76.9	76.9
1978	12.5		45	523	76.9	76.9
1979	12.5		60	52	83.2	83.2
1980	12.5		70	52	87.4	87.4
1981	12.5	20	70	52	90.3	90.3**
1982	12.5	20	75	52	91.9	91.9
1983			75	50	87.5	87.5
1984			75	45	86.3	86.3
1985			75	40	85.0	85.0
1986			75	35	83.8	83.8
1987			75	35	83.8	83.8
1988			75	35	83.8	83.8
1989			75	35	83.8	83.8†
1990			75	35	83.8	83.8
1991			75	34	83.5	83.5
1992			75	33	83.3	83.3‡
1993			50*	33	66.5	33
1994			50*	33	66.5	33
1995			50*	33	66.5	33
1996			50*	33	66.5	33
1997			50*	33	66.5	33
1998			50*	33	66.5	33
1999			50*	31	65.5	31
2000			50*	31	65.5	31
2001			50*	30	65.5	30
2002			50*	40	65.5	40

The U.K. had one of the *toughest* systems in the world here.

The *ringfence* is dropped for the PRT.
The boom begins.

The U.K. becomes famous for *lenient* terms. The boom continues.

* New fields receiving development approval after 16 March 1993 exempt from petroleum revenue tax (PRT). Also, these take statistics ignore the effect of uplifts on the PRT.

**The U.K. had one of the *toughest* systems in the world here.

† The *ringfence* is dropped for the PRT.

‡The U.K, becomes famous for *lenient* terms. The boom continues.

Government Take Is Not
Relevant in Some Important Situations

The government take statistic is almost totally meaningless in some extremely important Middle Eastern countries. In the *buy-backs* in Iran where the contracts focus on existing, well-known oil and gas fields, the take statistics are rarely mentioned. Depending on oil prices and other variables, the government take may be from 95% to 97% with these buy-backs. This sounds harsh, but that may or may not be the case, depending on the potential IRR that can be achieved. The risks are low relative to the harsh, cold risks associated with exploration. When exploration risk is missing, government take statistics move away from center stage, and discussions focus on IRR. This is also true of the proposed operating service agreement (OSA) in Kuwait for field rehabilitation and for the three massive projects contemplated in Saudi Arabia under the gas initiative tendered in December 2000 when 10 foreign oil companies were invited to express an interest in bidding.

In presentations last year to the Kuwaiti Parliament, the government take for the proposed OSA was quoted at 98%, and this was subsequently published in Kuwait newspapers. What was not published was that the statistic represented a *discounted* figure. Government share of DCF (discounted at 5%) was estimated at 98%. Government take statistics are usually quoted *undiscounted*. Government take always goes up when present value discounting is factored-in, and 5% discounting is not especially high. Yet, some Parliamentarians in Kuwait believe the proposed terms grant potential foreign contractors too high an IRR. Some Kuwaitis point to the kind of returns their government's investments obtain in the West and ask: "Why should Westerners get a higher IRR for their investments in Kuwait than we get in the West?" Similar discussions are taking place in Saudi Arabia.

With low-risk projects, this is where negotiations focus and it is appropriate. There is a lot at stake. What would be a reasonable ROR for a relatively well-defined, low-risk project where a company can put from $5 to 20 billion to work? Should the companies be allowed an IRR that is equal to or greater than their corporate cost of capital? Should companies be allowed an IRR that is greater than the IRR they receive for their exploration efforts? Most people agree that IRR for exploration generally speaking is not robust (as we discussed in the last chapter). Most agree that returns (or rather *potential* returns) should be lower where risk is lower.

These are complex issues. These are the central issues now in Saudi Arabia and Kuwait and with any government contemplating tendering *non-exploration* contracts. Additionally, these are the key issues where negotiations have stalled in Saudi Arabia. The Kingdom is contemplating rates of return of around 8–10% consistent with water and power projects in the Gulf and in Europe. The IOCs are reportedly demanding assured IRRs for the integrated projects as a whole. Some targeted IRRs quoted by the oil companies are as high as 15–20%. This is unrealistic. What is realistic, though, is that the negotiations are properly focused. Government take has little meaning here.

Examination of the imperfections associated with the government take metric provides important insight. Using any metric like this is more meaningful if both the strength and weaknesses are understood. Furthermore, it is not intended to serve as a stand-alone statistic.

While it is appropriate to discuss, measure, and negotiate using IRR with development or rehabilitation projects, it is not appropriate for exploration projects. It is impossible to predetermine an appropriate IRR for any given exploration project. This was discussed in my last chapter, particularly "Example 2: Bidding Terms" (*see* Chapter 13). When it comes to exploration projects, government take is extremely important. It is a function of prospectivity—there must be a balance. Heavy science goes into determining that balance, ranging from geoscience to political science.

If a balance can be determined and negotiators can agree upon basic terms, then a deal can be struck. A lot goes into this. There is important work to be done sorting out the other contract terms and language to ensure the document represents the deal as accurately as possible. There is nothing trivial about any aspect of this process.

REFERENCES

[1] From: "White Paper: A Proposal for Annotated Upstream Petroleum Regime Model Form Provisions," commissioned by the Organization of American States' SLA/OAS–CIDA Project: "International Business Transactions in the Americas: Legal Harmonization and Bijuralism," F. Alexander, February 26, 2002.

15

Additional Commentary on Key Issues

This chapter was developed in an effort to provide additional dimension to some of the key issues introduced in earlier chapters. The topics include:

- **The value of reserves in the ground**

- **ERR**

- **Booking Barrels**

- **Maximum Efficient (Production) Rate**

More on the Value of Reserves in the Ground

In the early 1980s, Triton Energy (now the Dallas Business Unit of Amerada Hess Corporation) made two giant discoveries in the Llanos region of Eastern Colombia. These discoveries, Cusiana and Cupiagua, were the largest discoveries in the western hemisphere for nearly two decades until they were eclipsed by deepwater discoveries in Brazil. Once the Triton discoveries were announced, the Triton stock price sky rocketed. This is understandable of course up to a point, but the stock market response went beyond that point. It appeared that Wall Street analysts and/or the market assumed the value of the reserves at the time of discovery (prior to development) to be around $3/BBL. The formula worked like this:

(1ˢᵗ) Recoverable reserves are estimated

(2ⁿᵈ) Triton's working interest share of reserves is multiplied by $3/BBL

(3ʳᵈ) This value is divided by the number of Triton common shares

(4ᵗʰ) Triton stock shoots up accordingly

The natural question is this: Was $3/BBL a reasonable value for undeveloped reserves in the Llanos? Drilling costs in this region were more than $25 MM per well. A very long and costly pipeline over the Andes Mountains was needed. It would take a long time to get production started. Country risk was heavy—the first drilling rig on location was burned down by rebels. And, on top of all of that, the fiscal terms were not great. Oil prices were still expected to be robust at that time (which was overoptimistic) but still—$3/BBL? Figure 15–1 was based on DCF analysis of the value of a possible discovery under various price, cost, and fiscal scenarios. Assuming an oil price on the order of $25–30/BBL, the value should have been more realistically around $1 to maybe $1.50/BBL—not $3/BBL. Conditions have to be very robust for undeveloped reserves to be worth that much. In fact, considering the conditions previously described, the reserves would have to be developed and producing before they would be worth anywhere near that much (*See* Figure 15–2).

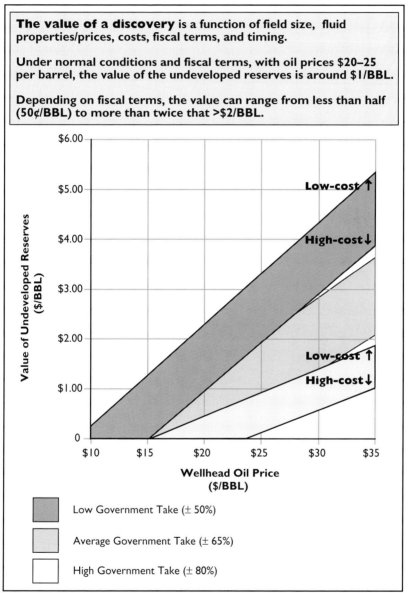

Fig. 15–1 Value estimate of newly discovered reserves

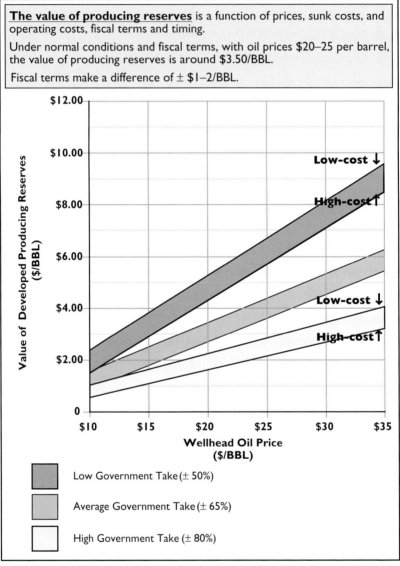

Fig. 15–2 Value estimate of producing reserves

There are typically around 100 discoveries reported worldwide each year. How much are these discoveries worth? In addition, there are billions of dollars of production acquisitions/sales each year. There is certainly a difference between the value of reserves at the point of discovery and developed-producing reserves. This chapter was generated to provide some insight into these reserve values.

There is unfortunately some confusion on this issue. The main source of confusion is that sometimes values are quoted in terms of working interest barrels and other times *entitlement* barrels. The following cash flow projection (Table 15–1) for a 100 MMBBL discovery illustrates where these different values can come from. A summary and analysis are provided in Table 15–2. Imagine a company makes a discovery expected to yield 100 MMBBLS of recoverable reserves and expects to develop the field in a fairly normal way—fairly normal costs, timing, and fiscal terms. The value to the company or contractor (assuming it holds 100% working interest) discounted at 12.5% comes to US $63.78 MM or roughly 64¢/BBL. However, company entitlement (cost oil + P/O) is 67.8% or 67.8 MMBBLS. The value per barrel for the *entitlement* barrels comes to 94¢/BBL.

Table 15–1 Cash Flow Projection for 100 MMBBL Discovery

Year	Annual Oil Production (MBBLS)	Oil Price ($/BBL)	Gross Revenues ($M)	Royalty 10% ($M)	Net Revenue ($M)	Capital Costs ($M)	Op. Costs ($M)	Depreciation ($M)	C/R C/F ($M)	Cost Recovery ($M)
	A	B	C	D	E	F	G	H	I	J
1	0	$20.00				30,000				
2	0	$20.00				80,000				
3	0	$20.00				150,000				
4	4,850	$20.00	97,000	9,700	87,300	90,000	17,275	70,000		58,200
5	9,420	$20.00	188,400	18,840	169,560		24,130	70,000	29,075	113,040
6	12,000	$20.00	240,000	24,000	216,000		28,000	70,000	10,165	108,165
7	10,200	$20.00	204,000	20,400	183,600		25,300	70,000		95,300
8	9,180	$20.00	183,600	18,360	165,240		23,770	70,000		93,770
9	8,033	$20.00	160,660	16,066	144,522		22,050			22,049
10	7,028	$20.00	140,580	14,058	126,522		20,544			20,543
11	6,150	$20.00	123,000	12,300	110,700		19,225			19,225
12	5,381	$20.00	107,620	10,762	96,858		18,072			18,072
13	4,709	$20.00	94,160	9,416	84,744		17,062			17,063
14	4,120	$20.00	82,400	8,240	74,160		16,180			16,180
15	3,605	$20.00	72,100	7,210	64,890		15,408			15,407
16	3,154	$20.00	63,080	6,308	56,772		14,731			14,713
17	2,760	$20.00	55,200	5,520	49,680		14,140			14,140
18	2,415	$20.00	48,300	4,830	43,470		13,623			13,623
19	2,113	$20.00	42,260	4,226	38,034		13,170			13,170
20	1,849	$20.00	36,980	3,698	33,282		12,774			12,770
21	1,618	$20.00	32,360	3,236	29,124		12,427			12,427
22	1,415	$20.00	28,300	2,830	25,470		12,123			12,123
Total	100,000		2,000,000	200,000	1,800,000	350,000	340,000	350,000		690,000

Year	Total Profit Oil ($M)	Gvt. Share ($M)	Company Share ($M)	Bonus ($M)	TLCF ($M)	Taxable Income ($M)	Income Tax 33% ($M)	Contractor Cash Flow	
								Undiscounted	12.5% DCF
	K	L	M	N	O	P	Q	R	S
1								(30,000)	(28,284)
2								(80,000)	(67,044)
3								(150,000)	(111,740)
4	29,100	11,640	17,460			(11,615)	0	(31,615)	(20,934)
5	56,520	22,608	33,912		11,615	41,207	13,598	109,224	64,288
6	107,835	43,134	64,701			74,866	24,706	120,160	62,867
7	88,300	35,320	52,980			52,980	17,483	105,497	49,062
8	71,470	28,588	42,882			42,882	14,151	98,731	40,814
9	122,545	49,018	73,527			73,527	24,264	49,263	18,102
10	105,979	42,391	63,587			63,587	20,984	42,603	13,915
11	91,475	36,590	54,885			54,885	18,112	36,773	10,676
12	78,787	31,515	47,272			47,272	15,600	31,672	8,174
13	67,682	27,073	40,609			40,609	13,401	27,208	6,242
14	57,980	23,192	34,788			34,788	11,480	23,308	4,753
15	49,483	19,793	29,690			29,690	9,798	19,892	3,606
16	42,041	16,816	25,225			25,225	8,324	16,900	2,723
17	35,540	14,216	21,324			21,324	7,037	14,287	2,046
18	29,848	11,939	17,909			17,909	5,910	11,999	1,527
19	24,865	9,946	14,919			14,919	4,923	9,996	1,131
20	20,509	8,203	12,305			12,305	4,061	8,244	829
21	16,697	6,679	10,018			10,018	3,306	6,712	600
22	13,348	5,339	8,009			8,009	2,643	5,366	426
Total	1,110,000	444,000	666,000				219,780	446,220	63,779

Table 15–1   Cash Flow Projection for 100 MMBBL Discovery [continued]

Year	Bonuses ($M)	Royalty 10% ($M)	Gvt. 40% Profit Oil ($M)	Income Tax 33% ($M)	Government Cash Flow ($M) Undiscounted	Government Cash Flow ($M) 12.5% DCF
	N	D	L	Q	T	U
1						
2						
3						
4		9,700	11,640	0	21,340	14,131
5		18,840	22,608	13,598	55,046	32,400
6		24,000	43,134	24,706	91,840	48,050
7		20,400	35,320	17,483	73,203	34,044
8		18,360	28,588	14,151	61,099	25,258
9		16,066	49,018	24,264	89,348	32,831
10		14,058	42,391	20,984	77,433	25,292
11		12,300	36,590	18,112	67,002	19,453
12		10,762	31,515	15,600	57,877	14,936
13		9,416	27,073	13,401	49,890	11,445
14		8,240	23,192	11,480	42,912	8,750
15		7,210	19,793	9,798	36,801	6,670
16		6,308	16,816	8,324	31,448	5,067
17		5,520	14,216	7,037	26,773	3,834
18		4,830	11,939	5,910	22,679	2,887
19		4,226	9,946	4,923	19,095	2,161
20		3,698	8,203	4,061	15,962	1,606
21		3,236	6,679	3,306	13,221	1,182
22		2,830	5,339	2,643	10,812	859
Total		200,000	444,000	219,780	863,780	290,855

A) **Production Profile** Thousands (M) barrels/year
B) **Crude Price**
C) **Gross Revenues** Thousands of dollars ($M)
D) **Royalty 10%** = (C * .10)
E) **Net Revenues** = (C – D)
F) **Capital Costs**
G) **Operating Costs** (Expensed)
H) **Depreciation** of Capital Costs (5-year SLD)
I) **Cost Recovery C/F** (if G + H + I > 60% of C)
J) **Cost Recovery** = (G + H + I) up to 60% of C

K) **Total Profit Oil** = (C – D – J)
L) **Government Share P/O 40%** = (K * .40)
M) **Contractor Share P/O 60%** = (K – L)
N) **Signature Bonus**
O) **TLCF** (See Column P)
P) **Taxable Income** = (C – D – G – H – L – N)
Q) **Income Tax (33%)** = [if P > 0, P * .33]
R) **Company Cash Flow** = (E – F – G – L – N – Q)
T) **Government Cash Flow** = (D + L + N + Q)

Table 15–2 Summary and Analysis of Table 15–1 Cash Flow Model

Simple PSC:	10% Royalty
	60% Cost recovery limit
	60/40% Profit oil split (in favor of the company)
	33% Income tax

Gross Revenues	$2,000,000	
Total Costs	-690,000	(34.5%)
Total Profit	1,310,000	
Bonus	- 0	
Royalties	- 200,000	
Government share profit oil	- 444,000	
Income tax	- 219,780	$863,780 (Gvt. Take)
Company cash flow	$446,220	
Company take		34% ($446,220/1,310,000)
Company entitlement		67.8% ($690,000 + 446,220)/2,000,000
Government take		**64%** ($863,780/1,310,000)

By the time these reserves are reported to shareholders according to U.S. SEC regulations, there will be even different values. This is because SEC requires disclosure of proved reserves only (discussed later in this chapter). When we in the industry discuss discoveries, the size of discoveries, development plans and required facilities we typically use *most likely* reserve numbers or P_{50} reserves. The P_{50} nomenclature comes from *probabilistic* reserve estimation methods such as Monte Carlo numerical simulation that yields a distribution of possible outcomes due to the uncertainty inherent in reserve estimates. The P_{50} reserve number represents that reserve estimate that has a 50% chance of being too high and a 50% chance of being too low. However, typical SEC reserve disclosure criteria requires P_{10} or *proved* reserves because of the greater level of confidence. The P_{10} category represents (theoretically) a reserve estimate that has a 90% chance of being too low and only a 10% chance of being overoptimistic.

RULES-OF-THUMB

U.S. Production Acquisitions

Reserve transactions worldwide represent billions of dollars each year as shown in Table 15–3. The United States is one of the most active markets for production transactions. For many years, the value of reserves in the ground in the United States has been said to be roughly *one third the posted price*. In fact this is not bad for a quick estimate and is borne out in Table 15–4. If someone expects to purchase say 5 MMBBLS of developed-producing reserves at a time when prices for that production are around $20/BBL, the value of the transaction would likely be around $33 MM [$20.00/3 * 5MM]. This kind of yardstick is not too accurate (±25%), but it provides a quick ballpark estimate. Reserves in the Gulf of Mexico typically trade for more than $1/BBL more than onshore production.

Table 15–3 Total Worldwide Reserve Deals

Year	Investment $MM	Added Reserves MMBOE	Reserve Cost $/BOE
1995	8,653	1,982	$4.37
1996	11,926	2,679	4.45
1997	37,632	8,724	4.31
1998	122,110	16,610	7.35
1999	96,498	16,201	5.96
2000	107,475	80,492	4.63
2001	81,532	64,709	4.13

From: John S. Herold Inc., *Oil & Gas Journal*, May 27, 2002, pg. 28

Table 15–4 U.S. Production Acquisitions
Greater than $1 MM and less than $400 MM

Year	Oil Price ($/BBL)	Reserves			Purchase Price $/BOE	Purchase Price as a % of Oil Price
		Oil MMBBLS	Gas BCF	MMBOE 10:1		
1979	$21.54	7	33	10	$6.81	32%
1980	$33.98	6	139	15	17.55	52
1981	$37.07	8	54	13	12.46	34
1982	$33.59	54	415	96	10.92	33
1983	$29.34	27	246	52	8.86	30
1984	$28.86	47	953	143	9.91	34
1985	$27.00	36	753	111	10.25	38
1986	$14.32	35	787	114	8.71	61
1987	$18.00	130	686	199	6.08	34
1988	$14.62	129	992	228	6.63	45
1989	$18.07	164	2,151	379	7.78	43
1990	$22.20	319	2,940	613	5.14	23
1991	$18.74	123	1,380	261	7.55	40
1992	$18.12	216	1,718	388	5.88	32
1993	$16.66	282	3,456	628	6.08	36
1994	$15.41	248	1,801	428	5.68	37
1995	$17.15	269	2,986	567	6.23	36
1996	$20.57	208	1,957	403	5.76	28
1997	$18.62	403	3,340	737	6.67	36
1998	$12.14	280	2,855	565	4.93	41
1999	$17.27	140	2,098	350	6.29	36
2000	$27.68	163	4,091	572	7.72	28
2001	$22.00	213	1,577	370	7.36	33
		3,506	**37,408**	**7,247**	**$6.59**	**35%**

From: Scotia Group database

International Production Acquisitions

For international production acquisitions, the DCF value of reserves has often been worth roughly one-half the posted price times contractor take. For example, in Indonesia where contractor take for much of the oil production is around 13%, the value of reserves in the ground with a wellhead price of around $20/BBL should be around $1.30/BBL [$20.00/2 * 0.13] discounted at 12.5%.

While the graph in Figure 15–2 was based on DCF analysis, the rule of thumb correlates fairly well. It is interesting, by the way, that this rule of thumb even worked in the early 1980s when higher discount rates were normally used. Back then too, though price estimates were often heavily overoptimistic, it balanced out.

This rule of thumb does not include any sunk costs. If a production acquisition is characterized by substantial sunk cost position, then the value of reserves can be much greater. This is because the purchaser would be able to recover those costs. Sunk cost positions, just like a *tax loss carry forward* (TLCF) can be valuable.

Strangely though, actual transaction values always seem to take place at much higher values. There are a number of reasons why. First, producing reserves often represent only part of the value of a production acquisition. Probable and possible reserves, non-producing (behind-pipe or undeveloped) reserves and undrilled prospects also add value. Also there is the ever-present *winners curse*. Just as companies overbid for exploration acreage, they overbid for production acquisitions. Tables 15–5, 15–6, 15–7, and 15–8 summarize various North American and worldwide statistics.

Table 15–5 1979–2001 U.S. Transactions

Transaction Size	Purchase Price	
	$/BOE 6:1	$/BOE 10:1
> $1 MM < $50 MM	$4.29	$5.73
> $1 MM < $100 MM	$4.29	$5.72
> $1 MM < $400 MM	$4.90	$6.59
From: Scotia Group database		

Table 15–6 U.S. Oil and Gas Reserve Activity, Cost

	Number of Transactions	Reserve Value $/BOE (6:1)	Reserve Value $/MCFE (6:1)	
2000				
1st Quarter	45	$4.61	81¢	
2nd Quarter	49	4.55		
3rd Quarter	34	5.20	Gas-dominated	
4th Quarter	28	5.10	deals	
2001				Median value for deals < $25MM is $4.05/BOE.
1st Quarter	39	$5.20	$1.14	
2nd Quarter	29	6.00		
3rd Quarter	25	7.15	Gas-dominated	Median value for deals
4th Quarter	29	5.88	deals	> $100 MM is $7.14/BOE.

From: Cornerstone Ventures LP, *Oil & Gas Journal*, May 27, 2002, pg. 34

Table 15–7 Worldwide reserve transactions 2001–2002

	2001	2002	
	Second Quarter	First Quarter	Second Quarter
Number of Transactions	40	34	42
Total Deal Value (Billions)	$22.81	$15.04	$13.97
Outside North America (Billions)	$6.5	$5.0	$8.5
Implied Reserve Value			
Worldwide ($/BOE)		$3.26	$3.30
US ($/BOE)		$6.05	$6.35
Canada ($/BOE)		$4.75	$7.87

From: "Upstream M&A Activity, Reserve Values Rebound in N. America", *International Oil Daily*, August 7, 2002, pg. 5

Table 15–8 Canadian Oil and Gas Reserve Activity, Cost

	Number of Transactions	Reserve Value $/BOE (6:1)	Reserve Value $/MCFE (6:1)
2000	87	CA$ 5.88 US$ 3.80	CA$ 1.19 US$ 0.77 Gas-dominated deals
2001	131	CA$ 8.52 US$ 5.54	CA$ 1.59 US$ 1.03 Gas-dominated deals
From: Cornerstone Ventures LP, *Oil & Gas Journal*, May 27, 2002 pg. 34. Conversion from CA$ to US$ based on US$0.65/CA$ 1.00			

More on Booking Barrels

After a couple decades of stability, oil prices went skyward with the 1973 embargo. Prices doubled, they doubled again, and then went higher. Suddenly, oil company balance sheets were obsolete. The true market value of a company's oil and gas properties was in no way reflected in the balance sheet. What good is that balance sheet? Nor were the dramatic increases in corporate value reflected in the income statement. Assets were *booked* at cost less depreciation. If a company made a billion barrel discovery, subsequent balance sheet and income statement entries would not look much different than how the company might instead have accounted for a dry hole. There was no way to adequately represent the actual *value* or change in value of oil and gas assets on company financial statements. An effort was made to change the way oil and gas assets were represented or reported.

Reserve Recognition Accounting [Birthplace of modern reserves disclosure]

In 1978, in response to a request from the United States SEC, the Financial Accounting Standards Board (FASB) announced a program of financial reporting, FAS 19, called Reserves Recognition

Accounting (RRA). The objective was to be able to recognize the value a company's proved reserves as an asset. Also, additions to proved reserves could be recognized as an asset, and the added value could be included in earnings.

The SEC originally intended RRA to replace full cost (FC) and successful efforts (SE) accounting methods. But RRA was only required as supplemental information during a trial period from January 1979 to November 1982. The FASB issued statement No. 25 in February 1979 suspending all but the disclosure requirements of FAS 19. It was determined that RRA could not replace FC and SE accounting due in particular to the inaccuracies and uncertainties of reserve estimates. In response to a request from the SEC, the FASB then developed disclosure requirements that were issued in November 1982 in statement No. 69 (FAS 69) "Disclosures about Oil and Gas-producing Activities."

RRA Chronology

SEC reserve reporting regulations began with the Federal Securities Laws, the Energy Policy and Conservation Act of 1975 and Rule 4-10(a) of Regulation S-X.

December 1977 FAS 19 issued

With this, the FASB adopted a form of SE accounting but also specified certain *supplemental data* including: (1) reserve quantities, (2) capitalized costs, (3) costs incurred.

August 1978   (before FAS 19 became effective) SEC issued ASR No. 253

This release adopted the SE accounting prescribed by FAS 19 and declared the intention to adopt the disclosure requirements if FAS 19 and the FC accounting method, which was subsequently done. This release also permitted the use of either SE or FC accounting methods for SEC reporting purposes. ASR No. 253 also required certain disclosure in addition to that recommended by FAS 19.

February 1979   FAS 25 issued

This suspended the effective date of FAS 19 as far as the accounting method to be used, but the disclosure requirements were not suspended.

February 1981   SEC issues ASR 289 *Financial Reporting by Oil and Gas Producers*

In this report, the SEC indicated it no longer considered RRA a potential method of accounting in primary financial statements. With this release, the SEC indicated its support of an undertaking by the FASB to develop a comprehensive package of disclosures for companies engaged in oil- and gas-producing activities.

November 1982 FAS 69 issued

These were the disclosure guidelines effectively commissioned by the SEC.

December 1982   SEC issues Reg. SK §229.302 adopting FAS 69

The SEC adopts FAS 69 to replace the SEC's own reporting requirements for oil- and gas-producing activities.

Summary of FAS 69 and related SEC Disclosures

The basic disclosure requirements are summarized as follows:

1. Method of accounting for oil and gas activities and manner of disposing of capitalized costs.

2. Publicly traded companies with significant oil and gas activity supplemental disclosure:

   a. Quantification of proved oil and gas reserves (reserves are further categorized as developed or undeveloped)

b. Capitalized costs associated with producing properties

c. Costs incurred for lease acquisition exploration and development activities

d. Results of operations for oil- and gas-producing activities

e. Standard Measure (SEC *Value* of Reserves) based on standardized DCF analysis of proved reserves

The basis of standardization is as follows:

a) Prices received at fiscal year-end for products sold

b) Prices are held constant, no escalation. Gas prices could be escalated if price escalation is included in gas purchase agreement

c) Costs are not escalated

d) A 10% discount rate is used

3. Interim financials should disclose information about major discoveries or other favorable or adverse events.

4. APB 20 on accounting policies calls for any significant accounting policy to be disclosed.

5. For enterprises following Reg. SX Rule 4-10 FC method, additional disclosures are required.

Costs associated with oil and gas E&P fall into four fundamental categories:

Lease Acquisition Costs—Costs associated with obtaining a lease or concession and rights to explore for and produce oil and gas.

Exploration Costs—Costs incurred in the exploration for oil and gas such as G&G, exploratory drilling, etc.

Development Costs—Costs associated with development of oil and gas reserves, such as drilling costs, storage and treatment facilities, etc.

Operating Costs—Costs required for lifting oil and gas to the surface, processing, transporting, etc.

Treatment of these costs is fairly straightforward. The one exception is exploration costs. This provides the basis for the two different accounting practices that are used in the industry; FC and SE. For all practical purposes, the FC accounting convention allows capitalizing (depreciating) unsuccessful exploration costs. Under SE accounting, unsuccessful exploration costs are expensed (i.e., not capitalized).

Summary of SEC Reserves Disclosure

The key criteria for determining whether or not reserves will be booked include:

- Right to extract
- Right to take in kind (not absolutely critical)
- Elements of risk and reward
- Economic interest

Reserves Recognition Governing Definitions

The SEC allows booking barrels if a mineral interest in a property exists. What constitutes a mineral interest is defined in SEC Regulations S-X Rule 4-10(b), SE Method,

(1) Mineral interests in properties. Including:

(i) fee ownership or a lease, concession or other interest representing the right to extract oil or gas, subject to such terms as may be imposed by the conveyance of that interest

(ii) royalty interest, production payments payable in oil or gas, and other non-operating interests by others

(iii) those agreements with foreign governments or authorities under which a reporting entity participates in the operation of the related properties or otherwise serves as *producer* of the underlying reserves (as opposed to being an independent purchaser, broker dealer, or importer); properties do not include other supply agreements or contracts that represent the right to purchase rather than extract oil or gas

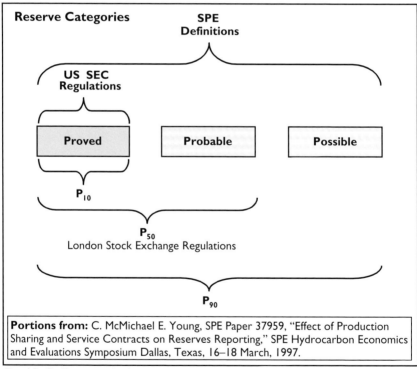

Fig. 15–3 *Reserve disclosure categories*

Booking Barrels

Booking barrels is based upon the concept of *reasonable certainty* and thus, the use in the United States of *proved* reserves only. Example restrictions that apply in regard to what constitutes proved reserves include:

- Only one legal (well) spacing from a discovery or producing well is allowed unless *continuity* can be demonstrated between two commercially producing wells. A well may have up to 8 offsets, which includes *diagonal* offset well locations.

- The definition of proved reserves is restricted to lowest known hydrocarbons (LKH).

- Either a successful well test or both a log and a core are required.

- Seismic hydrocarbon indicators such as amplitude anomalies (known as *bright spots*) that are often a strong indication of the presence of hydrocarbons cannot be booked unless they are drilled.

- Facilities must be in place for *proved* but not for *proved undeveloped* reserves.

The SEC posts latest interpretations on: www.sec.gov/divisions/corpfin/guidance/cfactfaq.htm#P279_57537

Booking Fuel

Produced gas consumed on-site for fuel in petroleum operations is being booked by companies. Presumably this would apply to oil consumed on-site if that were the case and with steam floods up to one third of the heavy oil production can be consumed as fuel to manufacture the steam needed for production if another source (such as gas) is not available.

Booking *imputed entitlement barrels*—taxes in lieu

Production-sharing systems with taxes *in lieu* where taxes are paid "for and on behalf of the Contractor" out of the NOC share of P/O are common: Egypt, Syria, Oman, Trinidad, Philippines, Qatar, etc.

Companies receive a lower entitlement in these systems than they otherwise would, had they paid the taxes directly. So companies are *grossing-up* the contractor share of P/O by dividing their P/O entitlement by (1 - tax rate) and they are booking this *imputed* entitlement.

For example, assume in Egypt the tax rate (being paid for and on behalf of the contractor) is 40% and contractor (actual) entitlement (of *proved reserves*) is 20 MMBBLS cost oil and 15 MMBBLS P/O for a total of 35 MMBBLS.

For *booking* purposes, the contractor would book the equivalent of 20 MMBBLs of cost oil + 25 MMBBLS *imputed* P/O [15 MM/(1 - 0.4)] or a total of 45 MMBBLS.

Booking BOE—Gas Conversion Factor

Some companies use a conversion factor based upon the *actual* heating value equivalent for the company's gas mix based on gas composition rather than the standard 6:1 ratio (6 MCF per Barrel heating equivalent). For example, a conversion factor of 5.75:1 (if that represented the relationship) would increase the BOE reserves by 4%. For a company with 10 TCF gas, the difference would be 72.5 MMBBLS.

Booking Royalty Oil

Some governments exercise their right to take their royalty in cash as opposed to in kind, which is their right under most contracts. When this occurs, the contractor or company will lift and sell this oil on behalf of the government, and some companies feel that there may be justification for booking these barrels as well.

Booking Gas Plant Liquids

Imagine a 100-MMCFD rich gas stream from a 1-TCF field (proved reserves) with a liquid yield of around 70 barrels per MMCF. Assuming the company holds 100% working interest in the gas field *and* the plant, there are two options. Converting the gas (at the field) using 6 MCF/BBL, the company would book 166.7 MMBOE. Alternatively, the company could book the products from the gas plant which would include the dry gas (after shrinkage) *and* liquids. Assuming 25% shrinkage, the company would book 195 MMBOE. [750 BCF @ 6:1 yields 125 MMBOE + 70 MMBBLS liquids].

Booking Barrels under Service Agreements

In countries that use service agreements where a company does not take title to hydrocarbons, the company will typically book the equivalent of its *economic* entitlement (or economic interest) as opposed to *lifting* entitlement (because there is no lifting entitlement). The economic entitlement would consist of the company deductions (or cost recovery) and pre-tax share of profits (the equivalent of P/O). Some governments are hyper-sensitive about companies booking barrels in their country. However, the problem can be avoided by booking the barrels but attributing them to a region not the specific country.

Fluctuating Entitlement

The reserves to which a company is entitled to *lift* (lifting entitlement) will often correspond to the reserves the company will *book* with of course the exception previously mentioned (Egyptian type arrangements with taxes in lieu) where P/O is *grossed-up*.

Under a typical PSC, the contractor entitlement is comprised of two components: cost oil and P/O. With fluctuations in oil (or gas) prices, the cost oil entitlement changes. With higher prices, it takes less oil for the contractor to recover costs. Thus, entitlement typically goes down with higher prices and up with lower prices. This frustrating phenomenon is not found with R/T systems.

Commentary on reserve value estimates and booking barrels

Reserves disclosure has improved since the early 1970s and most of that improvement occurred during the later part of that decade. But there has been relatively little change since the early 1980s. The time has come to provide additional information to shareholders. Proved reserves estimates, the heart and soul of U.S. reserves disclosure, are by design conservative. Yet companies make huge investment decisions based on something other than proved reserves. Following discovery, field development decisions are certainly based on more than just proved reserves. Typically companies use the *mean* of a distribution of possible reserve outcomes or what might be referred to as *most likely* recoverable reserves—something close to (usually) P_{50} reserves.

To develop an oilfield on the basis of proved reserves alone would be foolish. Production transactions too are based on more than just proved reserves. Yet shareholders who also make important investment decisions are limited to using proved reserves by the selective nature of modern reserves disclosure.

This reminds me of reserve certification work in the early part of my career where many consultants were proud of their *conservatism*. A company would hire a petroleum engineering firm to estimate recoverable reserves for a field and the firm would provide a *conservative* estimate. This implies they were capable of estimating what they actually thought the recoverable reserves were but chose to provide something less than that. The natural question was: *How conservative is the estimate?* If a company was going to make an important investment decision on the basis of a reserve estimate, then certainly it would be important to know just how conservative the consultant had been. If a company was taking the estimate to a banker as loan collateral, then the consultant's conservatism would likely be appreciated (mostly by the banker). But if the estimate was needed for a field development plan, then perhaps the same conservative attitude may not be best.

By definition, proved reserve estimates are conservative—typically P_{10}, the reserve estimate that has a 90% chance of being too low. So what use is that? To say that it is better than nothing is true, but that has a hollow ring.

Suppose a company were to make a significant discovery in Egypt. The natural question would be, *How much is this discovery worth to the company and shareholders?* Shareholders in the United States cannot rely on modern reserve disclosure to answer that question.

Furthermore, it could be many accounting periods following the discovery before shareholders and potential shareholders had even a modest (conservative) clue as to the significance of the discovery.

Modern reserve disclosure was borne out of an effort to quantify and provide information about the value of oil and gas assets. However, modern reserve disclosure in the United States falls short of that objective. In the meantime, while it is rather pathetic, a few rules-of-thumb may come in handy.

More on ERR

One area where I have found considerable discomfort and less-than-perfect-understanding is with the ERR calculation that is based on the assumption that costs by far outweigh revenues in a given accounting period. Again, the ERR represents the minimum share of revenues or production the government may expect in any given accounting period from royalties and/or P/O share. This statistic captures the royalty effect that results from the combination of cost recovery limit and subsequent P/O split. Both royalties and cost recovery limits (in conjunction with the P/O split) will guarantee the government a share of production in each and every accounting period.

The confusion surrounds the concept of *zero tax base* especially in an accounting period where a company receives P/O. The key is this: rarely is the company share of P/O equal to the tax base. A company can be in a no-tax-paying position even though it receives some P/O. A typical example from the cash flow model in the previous chapter is summarized here. The cash flow model (*see* Table 15–1) is for a 100 MMBBL field requiring $350 MM capital expenditures over a 4-year period. In Year 4, the first year of production, 4,850 MBBLS are produced, and $97 MM in gross revenues are generated. The tax calculation in that accounting period is as follows:

One approach

Gross Revenues		97,000 (Year 4)
Royalty (10%)	-	9,700
Net Revenues		87,300
Less Deductions		
Opex	-	17,275
Depreciation	-	70,000
Gvt. P/O	-	11,640
TLCF	-	0
Tax Base		(11,615)

Another approach

Company Revenues		
Cost oil		58,200 (Year 4)
P/O		+17,460 (Year 4)
Total Company Revenues		75,660
Less Deductions		
Opex	-	17,275
Depreciation	-	70,000
TLCF	-	0
Tax Base		(11,615)

The tax base is negative. There will be no taxes paid in this accounting period. The negative tax base is treated as a TLCF. In the next accounting period, it will be treated as an ordinary deduction. If there are sufficient revenues generated, this account will be depleted and the company will be in a tax paying position otherwise there will be an additional C/F balance at the beginning of the next accounting period. It works like this in almost all countries. The only exceptions are those rare PSCs where the tax base is *defined* as the company share of P/O.

Thus, in the example cash flow model (Table 15–1), in the first year of production, the system is *saturated*—the company hits the cost recovery limit and is in a no-tax-paying position. The ERR then is equal to 22%. This is because the government received $11.64 MM of P/O and $9.7 MM royalty oil, which represented 22% of the total gross revenues or production. Company AGR is equal to 78% which is about average worldwide [$$58.2 MM cost oil + $17.46 MM P/O out of $97 MM gross revenues].

More on MER

This section of Chapter 15 is a condensed summary of the book *Maximum Efficient Production Rate* by Daniel Johnston and David Johnston, published by the University of Dundee, Scotland (2002).

The natural resources of a nation, particularly the minerals, often constitute the lion's share of the nation's wealth. This is particularly true of the OPEC nations and other exporting countries. It is also true of many developing nations. But for other nations, it is true to various degrees. And in almost all countries, oil and gas production represents a disproportionately large share of the nation's budget. In Mexico for example, with around 3.5 million barrels per day of production, the petroleum sector (both upstream and down) represents around 15% of the gross national product (depending on oil prices), but it easily represents 40–50% of the nation's budget.

Most government officials view the mineral wealth of their nation as a gift from God. And, consistent with that view, waste in any form would be a sin. Furthermore, most of them as custodians of their nation's mineral wealth consider how their actions and decisions will be viewed by future generations. They believe their job is important—a sacred trust. The risk of *waste* looms large in their perspective.

Many believe considerable waste occurs when a reservoir is depleted too quickly, that it reduces recovery efficiency. Often too, terms such as *rape and pillage* or *reservoir damage* are used to refer to the results of what are considered to be aggressive oil reservoir depletion practices.

Thus the relationship between the rate-of-extraction from an oil or gas reservoir and ultimate recovery is an important subject. It receives a dramatic and emotional response from both the geotechnical and engineering ranks as well as politicians and citizens.

The central question is:

Do oil companies only care about short-term profits and produce fields too quickly thereby leaving behind more oil or gas than they should?

The objective in this chapter is to strip away the abundant emotional baggage and focus purely on technical and financial aspects of the relationship between how quickly oil and gas fields are produced and expected ultimate recovery (EUR).

The conclusions here are based on research collected as well as numerous formal discussions held on this subject over the past 20 years. There are 27 main references summarized and discussed in the book and the general conclusions are summarized as follows.

EXECUTIVE SUMMARY AND CONCLUSIONS

1. For conventional black-oil reservoirs comprising more than 90% of the world's 77 MMBOPD of production (2002), there is not a strong correlation between rate-of-extraction and ultimate recovery.

2. More than two-thirds of the 27 main summarized references in the study claimed ultimate recovery was either independent of extraction rates (or well spacing) or there was a positive correlation: faster extraction rates yield greater recovery (but not by much).

3. Three key factors influence the technical (non-financial) relationship between rate-of-extraction and ultimate recovery:
   - Drive mechanism
   - Reservoir petrophysical characteristics
   - Fluid properties

4. The general consensus regarding rate vs. recovery is summarized in Table 15–9:

Table 15–9 The Influence of Drive Mechanisms

Reservoir	Primary Drive Mechanism	General consensus regarding recovery
Oil	Solution Gas Drive	Independent of rate
	Water Drive	Rate sensitive—Higher production rates increase recovery but not by much.
	Gravity Segregation	Rate sensitive—Lower production rates increase recovery.
	Gas Cap Expansion	?
Gas	Water Drive	Rate sensitive—Higher production rates increase recovery.
	Volumetric Reservoirs	Independent of rate

5. For low permeability reservoirs with viscous oil, ultimate recovery is expected to improve with slower rates of extraction. These reservoirs, however, represent a fraction of world oil production.

6. The influence of timing due to time value of money is dramatic compared to modest differences in recovery if and where they exist relative to rate of extraction.

Discussion

The focus of this analysis is on the technical aspects of the relationship between rate-of-extraction and ultimate recovery. Yet when present value discounting (time value of money) is factored in, the subject takes on added dimension. Some authors state that there is clearly a correlation between rate-of-extraction and ultimate recovery because of the financial aspect that dictates when the field economic limit is reached. This is legitimate too, but sometimes confuses the issue. This analysis focused on the technical relationship between rate of extraction and ultimate recovery regarding primary drive mechanisms. There is little information on this subject as it regards secondary, tertiary, or enhanced recovery mechanisms. This study however, should provide valuable insight.

The opinion of many is that to produce oil or gas reserves slowly is like having money in the bank for future generations. However, the analogy is consistent with a bank interest rate of zero percent (0%). There is widespread but not universal agreement about the importance of the time value of money and where it fits. In addition to producing at higher rates, companies are accelerating production by starting up production or getting onstream faster. The financial benefits are substantial.

When it comes to the issue of how quickly reserves are produced once they come on-stream, the production to reserves ratio (P/R) metric provides excellent insight.

P/R is the percentage of total ultimate recoverable reserves for an oil or gas field produced in a peak year of production, and is expressed as a percentage.

$$P/R = \frac{\text{One year of production (BBLS) [in a peak year]}}{\text{Expected ultimate recoverable reserves (BBLS)}}$$

For example, an oilfield with 100 MMBBLS of recoverable reserves producing at 40,000 BOPD (14.6 MMBBLS/year) has a P/R ratio of 14.6%.

$$P/R = \frac{40,000 \text{ BOPD } / 365 \text{ days}}{100 \text{ MMBBLS}} = \frac{14.6 \text{ MMBBLS}}{100 \text{ MMBBLS}} = 14.6\%$$

P/R ratios around the world	Typical ranges—P/R Ratio
Conventional oilfields	8–25%
Small oilfields	15% or more
Medium-sized oilfields (50–200 MMBBLS)	10 to 12%
Indonesian oilfields	20–25%
Deepwater developments	20% or more
Gas fields worldwide	2–8% (if they are producing at all)
Gulf of Mexico gas wells	40%

Typically production decline rates following the plateau (peak) rate of production are close to or greater than the P/R ratio. For example, in Indonesia where P/R ratios can be on the order of 20–25%, the production decline rate is often more than 25%. These P/R statistics are the practical result of development decisions that weighed geotechnical and financial elements in the context of development alternatives in such a way as to maximize present value.

One of the key variables in this analysis is the well spacing. Therefore our research targeted anything that dealt with rate of extraction or well spacing and the effect on ultimate recovery.

Oil companies are developing fields these days more quickly and at faster rates. The time from discovery to first production is not as long as it once was. These lead times are nearly one-fourth what they once were. There are a couple of reasons for this. First of all, the basins of this world are more mature, and infrastructure is better established. Also, companies are more sensitive to the present value timing effects and the benefits of getting production onstream more quickly.

Figure 15–4 provides a stylized illustration of the stages in the life cycle of a typical oilfield or drilling center such as an offshore platform. Larger fields or fields that are developed in *stages* such as multi-platform developments would have numerous such profiles— each associated with a separate drilling center or development stage.

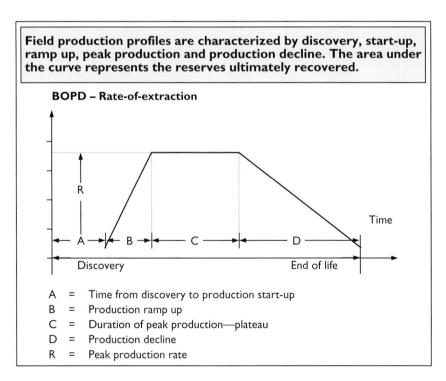

Fig. 15–4 Typical oilfield life cycle

Figure 15–5 shows the same thing but with an actual example from the Murcheson North Sea field developed in the early 1980s.

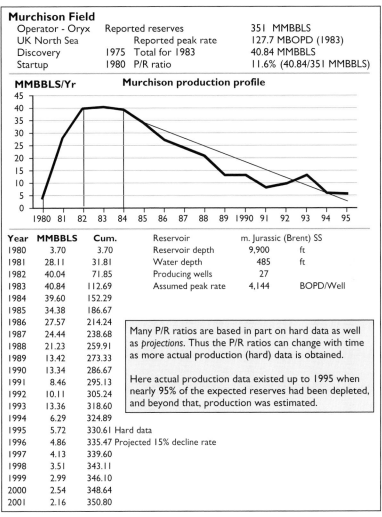

Murchison Field

Operator - Oryx	Reported reserves	351 MMBBLS
UK North Sea	Reported peak rate	127.7 MBOPD (1983)
Discovery	1975 Total for 1983	40.84 MMBBLS
Startup	1980 P/R ratio	11.6% (40.84/351 MMBBLS)

Murchison production profile — MMBBLS/Yr

Year	MMBBLS	Cum.
1980	3.70	3.70
1981	28.11	31.81
1982	40.04	71.85
1983	40.84	112.69
1984	39.60	152.29
1985	34.38	186.67
1986	27.57	214.24
1987	24.44	238.68
1988	21.23	259.91
1989	13.42	273.33
1990	13.34	286.67
1991	8.46	295.13
1992	10.11	305.24
1993	13.36	318.60
1994	6.29	324.89
1995	5.72	330.61 Hard data
1996	4.86	335.47 Projected 15% decline rate
1997	4.13	339.60
1998	3.51	343.11
1999	2.99	346.10
2000	2.54	348.64
2001	2.16	350.80

Reservoir	m. Jurassic (Brent) SS	
Reservoir depth	9,900	ft
Water depth	485	ft
Producing wells	27	
Assumed peak rate	4,144	BOPD/Well

Many P/R ratios are based in part on hard data as well as *projections*. Thus the P/R ratios can change with time as more actual production (hard) data is obtained.

Here actual production data existed up to 1995 when nearly 95% of the expected reserves had been depleted, and beyond that, production was estimated.

Fig. 15–5 Murcheson oilfield profile

One of the most important variables influencing rate-of-extraction is the number of wells a company is willing to drill. The more wells, the faster the rate of extraction. Submersible pumps can also increase production rates. This discussion focuses on the key variable: number of wells.

Figure 15–6 depicts two development scenarios for a single fault block: one with two wells and one with six wells. A structure map is shown of a reservoir at around 8200 feet with dip closure to the West and North and fault bound to the Southeast. By placing wells on the crest of the structure, either scenario could deplete the reservoir— except scenario #1 will take longer.

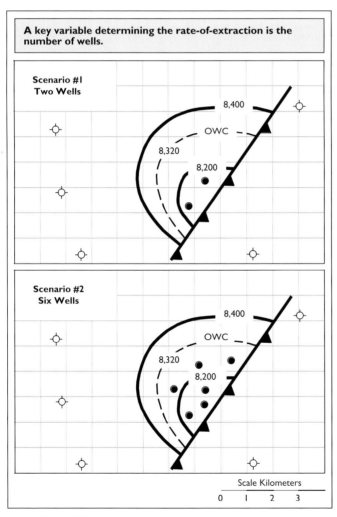

A key variable determining the rate-of-extraction is the number of wells.

Scenario #1
Two Wells

8,400

OWC

8,320

8,200

Scenario #2
Six Wells

8,400

OWC

8,320

8,200

Scale Kilometers

0   1   2   3

Fig. 15–6 Effect of well spacing

The two-well scenario has the wells located on the crest of the structure and, theoretically, this could produce all of the recoverable reserves. But it might take a long time.

If anything, the recovery factor may be greater due to improved sweep efficiency with the additional wells.

While there were 27 main references included in this study, this represents various other authors and references.

Much of what we observed is captured by Vietti (as follows) in 1945, but we believe there has been progress—the difference of opinion is perhaps not so wide now.

> *That such a wide difference of opinion as to the effect of well spacing upon recovery should exist among the well informed is particularly disturbing to management in the petroleum industry and to the layman.*

From: Vietti, V., Mullane, J., Thornton, O., Everdingen, A., *The relation between well spacing and recovery*, presented at 25th Annual SPE Meeting, Chicago, IL, 1945, pp. 160–174.

Table 5–10 summarizes the general conclusions and/or findings of the 27 main references.

Table 15–10 Summary of References

	Author(s)	Date	General Conclusionsre: Correlation (1)			
1	**Vietti et al**	1945		I		
2	**Craze and Buckley**	1945		I		
3	**Elkins**	1946		I		Well spacing
4	**Campbell**	1957				Inconclusive—general statements.
5	**Buckwalter et al**	1958	P			Water flood
6	**Craft and Hawkins**	1959		I	N	Solution drive—independent. Low permeability and gravity segregation—rate dependent.
7	**Cole**	1961	P	I		Volumetric gas reservoirs—independent Water drivegas reservoirs—various.
8	**Heuer et al**	1961		I		Well spacing and production rates.
9	**Arps**	1968				Issue not mentioned in detailed study!
10	**Cook and Fulton**	1973		I		Much of this deals with coning
11	**Miller and Roger**	1973	P			Water drive
12	**Lee et al**	1974	P			Water drive modeling study
13	**Savage et al**	1974			N	Simulation study
14	**Beveridge et al**	1974	P			Various other studies quoted
15	**Stright et al**	1975	P			Water flood studies
16	**Langham**	1978	P			
17	**Gerov and Mincheeva**	1978		I		
18	**Jordan et al**	1978		I	N	Secondary water floods
19	**Everdingen and Criss**	1980	P			Denser well spacing—higher recovery.
20	**Banks**	1980			N	
	Banks	1998			N	
21	**Lohrenz et al**	1980		I		
	Lohrenz et al	1980		I		
	Lohrenz et al	1981	P		N	Water drive and Gravity Segregation
22	**Muscat**	1981	P	I		Well spacing
23	**El-Khatib**	1982	P		N	Faster can be better except with *low formation capacities.*
24	**Allen and Seba**	1993		I		
25	**Hyne**	1995			N	
	Hyne	1997			N	
26	**Anonymous**	1997		I	N	Rate insensitive up to 15% P/R – Minagish.
27	**C-NOP Board**	2000	P			Hybernia – Slight increase at higher P/R.

(1) Correlation between rate-of-extraction and ultimate recovery:

P = Positive (faster rate-of-extraction yields greater recovery)
I = Independent (ultimate recovery is independent of rate-of-extraction)
N = Negative (faster rate-of-extraction reduces recovery)

16

Sample Contracts & Summaries

EXAMPLE CONTRACTS

The example contracts in this chapter are selected from information in the public domain. They should provide perspective on the rich diversity that exists and some of the key aspects of the commercial terms of contracts between governments and oil companies. Every effort has been made to ensure the accuracy of the data but this is a difficult chore. Key criteria for including a particular contract or system was if sufficient information was available to calculate government take and the ERR.

Note: Uncertainty or lack of information is indicated with a question mark (?)

ANGOLA

Offshore mid-1990s PSC

Area	4,000–5,000 km² (1–1.2 MM acres)
Duration	**Exploration** 3 years + 1 + 1 + .5 + .5
	4 years + 2 for deepwater
	Production 20 years from date of discovery
Relinquishment	All except development areas after 5 years onshore
	after 6 years deepwater
Exploration	
Obligations	**Conoco** 1986 $60 MM 4,000 km Seismic + 6 wells
Negotiable	**Total** 1989 $9 MM Seismic + 2 wells
Signature Bonus	
Rentals	$300/km² for development areas
Royalty	None
Cost Recovery	50% limit
	40% Uplift on development costs
Depreciation	Exploration costs expensed
	Development costs 5 year straight line (was 4 years)

Profit Oil Split	**MBOPD**	**Company**
(Typical)	0–25	50–60%
	25–50	30
	50–100	20
	> 100	10

Taxation	In lieu—paid by Sonangol (50%)
	With economic equilibrium/stability clause
Ringfencing	For cost recovery
	Around license for exploration
	Around field for development
DMO	Pro rata option/right up to 40% of production
Gvt. Participation	Up to 51% in early contracts (assumed here)
	After 1997, typically 20% *Heads up*
Other	Price cap formula Gvt. takes 100% above $32/BBL (1991)

Government Take				Effective Royalty Rate	Lifting	Savings Index	Data Quality
Downside	**Mid-range**	**Upside**	**Margin**				
78%	76%	77%	78%	31%	64%	35¢	Good

AZERBAIJAN

AIOC PSC I 20 Sept., 1994, Amoco, Lukoil, McDermot, et al.

Area	Azeri & Chirag & Deepwater Gunashli
Duration	"Basic Term" 30 years
Exploration	An extension of basic term provision for discovery
Relinquishment	No—not in the normal sense
Obligations	Early Production
	3-D Seismic over entire area (20,000 km² full fold)
	3 appraisal wells + Environmental baseline survey
Bonuses	$300 MM—3 installments less 10% SOCAR WI if back-in option exercised
	1/2 [$150 MM] = Signature Bonus; 1/4 [$75 MM] @ 40,000 BOPD;
	1/4 [$75 MM] when oil exported from MEP (Main Export Pipeline)
Royalty Rentals	None
Cost Recovery	No limit on OPEX
	CAPEX limited to 50% of remainder "all finance costs" recoverable
Depreciation	4 years for Equipment and capital assets
	Abandonment Costs—10% of CAPEX when reserves 70% depleted (UOP)

Profit Oil Split

	Early & <$3/BBL & >$4/BBL			Late & <$3/BBL & >$4/BBL	
RROR	**P/O Split**	**P/O Split**	**P/O Split**	**P/O Split**	**P/O Split**
< 16.75%	30/70%	25/75%	25/75%	25/75%	20/80%
16.75-22.75%	55/45%	50/50%	55/45%	55/45%	50/50%
> 22.75%	80/20%	75/25%	80/20%	80/20%	75/25%

Gas Clause—exclusive right to negotiate

Taxation	25% Profit Tax (in later contracts paid by SOCAR)
	0% VAT—5% Withholding on Subs (25% of assumed 20% profit)
Ringfencing	Yes
DMO	10% @ Market price at delivery point + 10% @ Market price at MEP
Gvt. Participation	10% Government pays for costs between Execution & Effective date
	[LIBOR + 4%]
	SOCAR use of Chirag I Platform valuation adjustment
G&A	Expensed—1st $15 MM 5%; 2nd $15 MM 2%; > $30 MM 1%;
Other	Opex @ 1.5%
	Hiring Quotas: Employee/Expat Wage Tax

Government Take				Effective Royalty Rate	Lifting	Savings Index	Data Quality	
								* Wide range
Downside	**Mid-range**	**Upside**	**Margin**					
38%	64%	70%	80%	0%	59%	41¢ *	Good	

CHINA

Offshore PSC 1990s

Duration	30 years	**Exploration** 7 years
		Production 15 years + extensions with approval
Relinquishment	25% after Phase I, 25% of remaining after Phase II,	
	All remaining at end of Phase III excluding development areas.	
Bonuses	(?)	

Royalty	**BOPD**	**Royalty**	**MMCFD**	**Royalty**
	Up to 20,000	0%	Up to 195	0%
	20,001–30,000	4%	195–338	1
	30,001–40,000	6%	338–484	2
	40,001– 60,000	8%	>484	3
	60,001– 80,000	10%		
	> 80,001	12.5%		
	(Tons/yr converted to BOPD at 7:1)			
	(MM m3/yr converted to MMCFD at 35.3:1)			

Pseudo Royalty	5% VAT (was Consolidated Industrial & Commercial Tax based on gross revenues)
	(VAT is 13% for Chinese companies)

Profit Oil Split	**BOPD**	**Gvt/Contractor**
(Negotiable) Example	Up to 10,000	10/90% *
	10,000–20,000	20/80%
	20,000–40,000	30/70%
	40,000–60,000	40/60%
	60,000–100,000	50/50%
	>100,000	60/40%
	* *Some contracts start at 95%*	*(X factor) and slide to 45%.*

Cost Recovery Limit	50%–62.5% With 9% interest cost recovery on development costs
	All costs expensed for cost recovery purposes

Taxation	30% Income Tax (15% in Hainan Province)
	3% Local Income Tax
	10% Surtax
	Contractors must also pay vehicle and vessel usage, license tax and individual income tax.

Depreciation	6 Year SLD for Development costs, Exploration costs expensed
Ringfencing	Yes for cost recovery but not for taxes;
	offshore costs are deductible against onshore income
Gvt. Participation	Up to 51% upon Commercial Discovery
	No repayment of past exploration costs.

Government Take				Effective Royalty Rate	Lifting	Savings Index	Data Quality	There is a wide range of terms that have been negotiated in China
Downside	**Mid-range**	**Upside**	**Margin**					
76%	75%	74%	73%	13%	85%	51¢	Good	

Other sources indicate Gvt.
Take ranging as low as 64%.

COLOMBIA

Association Contract post-1994

Area	Various Blocks
Duration	28 years including exploration period
	Up to 10 years for exploration
Relinquishment	50% at end of year 6
	25% more at end of year 8
	Remainder dropped after Year 10 except commercial field & 5 km band.
Exploration Obligations	Yes—negotiated
Bonuses	
Royalty	20%
	1990 War Tax 600–900 pesos/BBL—roughly $1/BBL for
	1st 6 years of production. Phased out in 1995–1997 (?)
	Cusiana/Cupiagua only(?)
Taxation:	35% Income Tax
	Remittance Tax 12%, 10% for 1997, 7% after 1997
Depreciation	Assumed 5 yr SLD for Exploration, 7 yr for Development
Ringfencing	Yes—Effectively
Gvt. Participation	50% Carried Interest
	At back-in Ecopetrol reimburses 50% of cost of successful wells.

Government Working Interest R factor based:

			Government Share	
		Successful		
R	**Net**	**Exploration**	**Development**	**Operating**
Factor	**Production***	**Costs**	**Costs**	**Costs**
< 1	50%	50%	50%	50%
1 - 2	1-(50%/R)	50%	50%	1-(50%/R)
> 2	75%	50%	50%	75%

* *Not including royalty*

If R is < 1.8 or cumulative production < 60 MMBBLS Gvt. production share stays at 50%.

Government Take				Effective Royalty Rate	Lifting	Savings Index	Data Quality
Downside	**Mid-range**	**Upside**	**Margin**				
87%	83%	82%	81%	20%	<80%	60¢	Good

COLOMBIA

TIBU Incremental Production Contract (IPC) 1998±
Ecopetrol Operator—Halliburton and Others

Area	On international border with Venezuela
Duration	Pilot Phase 2 years
	Development Phase 18 years + 10 yr extension
	30° API for Tertiary Barco; 45° API Cretaceous Catatumbo
	Some Exploration potential
Relinquishment	(?)
Obligations	Pilot Phase $15 MM (Bid Item - $10 MM minimum)
	Non-reimbursable—$1 MM Guarantee
	Development Phase $80 MM
Bonuses	No
Royalty	20%
Base Production	3,000 BOPD—4% Decline—Ecopetrol
	Ecopetrol pays 100% for base production. Costs shared
Incremental Production	73.5%/26.5% Halliburton/Ecopetrol
Taxation:	35% Income Tax
	7% Remittance Tax
Ringfencing	Yes—Effectively
Gvt. Participation	Government Share: Incremental Production R factor based:

R Factor	Government Share Production*	
< 1	26.5%	* *Not Including Royalty*
1 – 2	Slides from 26.5 to 50%	
> 2	50%	

If R is < 1.0 and cumulative Production < 30 MMBBLS
Gvt. Production share stays at 26.5%.

Government Take				Effective Royalty Rate	Lifting	Savings Index	Data Quality
Downside	**Mid-range**	**Upside**	**Margin**				
76%	72%	73%	75%	20%	<80%	61¢	Good

CONGO Br.

PSC 1994 New Hydrocarbon Law

Duration	Exploration	4 + 3 + 3 years
	Production	20 + 5 years
Relinquishment	50% after 4; 50% after 3 more	
Obligations	Typically Seismic + 4 wells in 1st period	
Signature Bonuses	Yes—not cost recoverable or tax deductible	
Royalty	15% Official for Oil (Negotiated)	
	Gas is negotiable	
Cost Recovery Limit	50-60% (may be 70% in deepwater)	
	Portion of excess cost oil goes to Government	

Profit Oil Split (Negotiable)	Example Splits	
	BOPD	**Gvt./Contractor**
	Up to 20,000	30/70% 40/60%
	20,000–40,000	50/50 50/50
	> 40,000	70/30 60/40
	Some may be straight 50/50% split;	
	May be linked to "realized" prices	

Taxation	35% for 1st 5 years sliding up to negotiated level
	Contractor exempt from other taxes except registration, stamp duty and service taxes.
Depreciation	Exploration Costs expensed—Others 20% SLD
DMO	May be required—pro rata—full market price
Ringfencing	Development area one ring fence, Minister has discretion over widening
Gvt. Participation	Negotiable (10–15%) (assumed 15%)
	One recent contract 20% carry through Expl. and Development!
Other	Price Cap formula,
	Gvt. Profit Oil increases to 82–85% if price goes above $22/BBL.

Government Take				Effective Royalty Rate	Lifting	Savings Index	Data Quality
Downside	Mid-range	Upside	Margin				
78%	73%	71%	70%	33%	64%	37¢	Good

287

COTE D'IVOIRE

27 June, 1995 Block CI-11 PSC Pluspetrol

Area	335,179 acres
Exploration Obligations	Phase I 1.5 yrs 1 well; $4 MM
	Phase II 2 yrs 2 wells; $8 MM
	Phase III 2 yrs 2 wells; $8 MM
Duration	
Appraisal	2 years for oil discovery 4 for gas + 6 month ext.
Exploitation	25 years + option to extend 10 years
Relinquishment	25% of original after Phase II & Phase III
Bonuses	Signature $300K in vehicles and office equipment
	Production $1, 3, 5, & $10MM @ 10, 20, 30 & 50 MBOPD Gas 6:1
Royalty	None
Cost Recovery Limit	40% All costs expensed
	[75% of interest costs and fees are recoverable,
	Bonuses not cost recoverable]

Profit Oil Split

Production * MBOPD	Contractor Share	Production MMCFD (Qtr)	Contractor Share
Up to 10	40%	Up to 75	40%
10–20	30	75 - 150	30
20–30	20	Over 150	(?)
Over 30	10		

* *Avg. production rate during quarter*

Taxation	Paid by Nat. Oil Co. on behalf of contractor (50%)
Ringfencing	Yes
DMO	10% of Contractors crude at 75% of market price
Gvt. Participation	Around B1-8X 40% +
	outside special area Gvt. carried through exploration 10% + option to purchase 10%
Other	75% Minimum Employment Quota: $100K/yr training > $150K/yr
G&A	During expl/appraisal 4% of costs
	Development 3% up to $3MM: 2.5%
	$3–6MM: 1.5%> $6MM

Government Take				Effective Royalty Rate	Lifting	Savings Index	Data Quality
Downside	**Mid-range**	**Upside**	**Margin**				
86%	74%	73%	73%	36%	55%	34¢	Good

ECUADOR

7th Round PSCs 1995

Area	Maximum 200,000 hectares onshore, 400,000 offshore
Duration	**Exploration** 4 + 2 years for Oil; 5 + 2 for Gas
	Production 20 years for Oil; 25 years for Gas
Relinquishment	No interim relinquishment
Exploration Obligations	$12–16 MM 0–1 wells Amazon
	$ 8-13 MM 1–2 wells W. Coast Region
Bonuses	Various fees
Rentals	None
Other	$100,000 Bidding Fee Amazon, $50,000 W. Coast
Royalty	No Royalty - paid out of National Oil Co. share

Gross Production Split	**BOPD**	**Contractor/Gvt**
(Example)	< 25,000	75/25%
	25,000 - 50,000	65/35
	> 50,000	50/50
	Typical ranges for 1st tranche from 90/10% to 65/35%	
	Contractor gets added 2% for each °API below 25° API	
	Gvt. gets added 1% for each °API above 25° API	

Taxation	15% employees statutory profit sharing deductible for income tax
	25% income tax
	36.25% effective rate
	Ad Valorem (Total Assets) $0.1 + 0.15 \approx 0.25\%$
Depreciation	Tangible costs pre-production 5 year SLD, Post UOP
	There is a limit on G&A = 15% of Exploration investment
	No "financial cost" recovery for Exploration
	Limit on payments to Home Office 5% of taxable base (?)
Ringfencing	Yes—Around contract area
DMO	Possible—pro rata
Gvt. Participation	None under PSC or predecessor RSA; Pre1982 was 25%
Other	50¢/BBL, Sept. 1997 environmental tax decree on production
	25¢/BBL transportation (previous contracts exempt)

Government Take				**Effective Royalty Rate**	**Lifting**	**Savings Index**	**Data Quality**
Downside	**Mid-range**	**Upside**	**Margin**				
75%	63%	57%	54%	26%	73%	63¢	Good

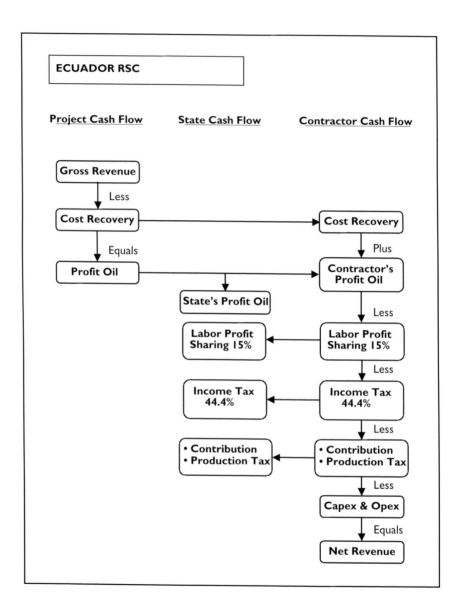

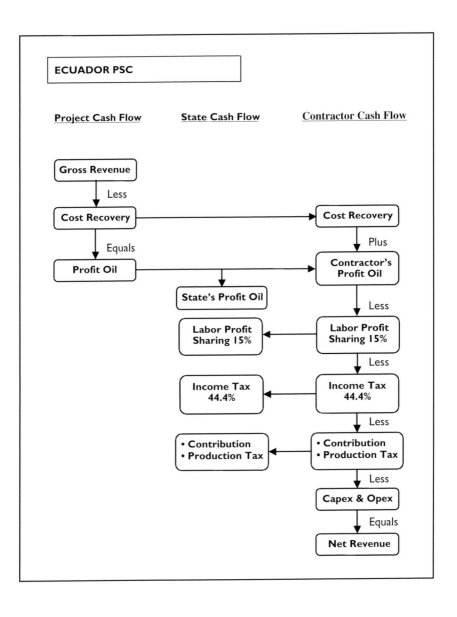

ECUADOR PSC

Project Cash Flow   **State Cash Flow**   **Contractor Cash Flow**

Gross Revenue

Less

Cost Recovery ──────────────→ Cost Recovery

Equals   Plus

Profit Oil ──────────────→ Contractor's Profit Oil

State's Profit Oil   Less

Labor Profit Sharing 15% ←── Labor Profit Sharing 15%

Less

Income Tax 44.4% ←── Income Tax 44.4%

Less

• Contribution
• Production Tax ←── • Contribution
• Production Tax

Less

Capex & Opex

Equals

Net Revenue

EGYPT

Norsk Hydro - Kufpec -- Ras El Hekma PSC

Area	Ras El Hekma Block W. Desert 535 km²
Duration	**Exploration** 3 phases 3+2+2 years
	1 well + 300 km 2-D, + 1 well, + 2 wells
	with performance bonds
	Production 20 years from discovery not to exceed
	35 years oil and gas
Relinquishment	25% + 25% @ 3 & 2
	If no commercial production after 4 years of discovery—100%
Exploration Obligations	3 phases 3 + 2 + 2 years/$4.5 + $3.5 + $7 MM
	1 well + 300 km 2-D + 1 well + 2 wells
Signature Bonus	$300K on Effective Date not cost recoverable
Production Bonuses	$1MM & 2MM @ 10 & 20 MBOPD
Royalty	10% "borne and paid by EGPC"
Rentals	BOE = MSCF*h*0.136 h = BTU/MCF
Cost Recovery	40% Excess cost oil 100% to EGPC
Depreciation	25%/year expl. & dev. costs
Profit Oil Split	**MBOPD** **Suez**
(in favor of Government)	Up to 10 70/30%
	10–25 75/25
	> 25 80/20
	Gas & LPGs 78/22%
Profit Gas Split	Typically negotiated
Taxation	Paid by EGPC
Ringfencing	Yes
Other	Gas price—15% discount from HSFO FOB Med

Government Take				Effective Royalty Rate	Lifting	Savings Index	Data Quality
Downside	Mid-range	Upside	Margin				
98%	74%	75%	76%	44%	47%	26¢– $1.00	Good

EGYPT

Alliance Intl., 1997 PSC

Area	Block G Central Sinai	4.5 MM Acres Onshore
Duration	**Exploration**	3 phases 3+2+2 years
		1st phase $6MM 4 wells + 2-D, and 3-D
		with $6 MM performance bond
	Production	20 years from discovery not to exceed
		35 years oil and gas (assumed)

Relinquishment	25% + 25%
Exploration Obligations	4 wells + 2-D, and 3-D
	$6 MM for the initial 3 year term

Signature Bonus	$1 MM
Production Bonuses	$2 MM @ 25,000 BOPD
	$4 MM @ 50,000 BOPD
	$8 MM @ 75,000 BOPD

Royalty	"borne and paid by EGPC"
Cost Recovery	35%
Depreciation	25%/year expl. & dev. costs (assumed)

Profit Oil Split	**MBOPD**	**Split**	
(in favor of Government)	Up to 5	74/26%	Some newer contracts
	5–10	76/24	start at 70/30%.
	10–20	78/22	
	20–40	81/19	
	40–50	83/17	
	> 50	85/15	

Profit Gas Split	Negotiated
Taxation	In lieu (the rate is 40.55% for petroleum companies)
Ringfencing	Yes
Gvt. Participation	None
Other	Gas price—15% discount from HSFO FOB Med

Government Take				Effective Royalty Rate	Lifting	Savings Index	Data Quality
Downside	**Mid-range**	**Upside**	**Margin**				
100%	76%	75%	75%	49%	49%	24¢	Good

EQUATORIAL GUINEA

United Meridian/Conoco PSC, 1992

Area	Offshore Blocks A-14, C-16, B-14, B-15, C-14, C-15
Duration	**Exploration** 1 + 3 + 2 years
	Production
Relinquishment	
Exploration Obligations	450 km Seismic 1st year
Bonuses	Signature $0.5 MM
	Discovery $1 MM
	First Production $2 MM
	50,000 BOPD $5 MM
Royalty	10% [Some reports of a 25 MMBBL holiday(?)]
Cost Recovery Limit	100% Assumed all costs expensed for cost recovery

Profit Oil Split	**Water Depth**		
"Net Crude Oil"	**Contractor's**	< 200 m	> 200 m
after Royalties and Cost Recovery	**Pre-Tax**	Contractor	Contractor
	Real ROR *	Share	Share
	Up to 30%	100%	100%
	30–40	75	80
	40–50	40	60
	> 50	20	40 * USCPI

$FSA(Y) = FSA(Y\text{-}1) * (1 + .30 + i) + NCF(Y)$ *[First Share Account]*

$SSA(Y) = SSA(Y\text{-}1) * (1 + .40 + i) + [NCF(Y) \text{-} GS1(Y)]$ *[Second Share Account]*

Taxation	25%
Depreciation	Exploration costs expensed
	Development costs 4-5 yr SLD
Ringfencing	Yes for cost recovery no for tax purposes
DMO	If requested, a portion of net crude oil at market prices.
Gvt. Participation	5%

	Government Take				Effective Royalty Rate	Lifting	Savings Index	Data Quality
Example	Downside	Mid-range	Upside	Margin				
<200m	42%	47%	74%	87%	10%	80%	75¢	Good
>200m	43%	45%	65%	73%	10%	83%	75¢	

GUATEMALA

PSC 1997

Duration	**Exploration** 3 + 2 + 1
	Production
Relinquishment	30% + 20% (original areas) or 50% by year 5
Exploration Obligations	
Bonuses	Signature, bonus and sliding scale training fees
Rentals	
Royalty	20% ± 1% for every degree above/below 30°API
	5% minimum below 15° API
	5% on gas and condensate
	35% for pre Commerciality production
Cost Recovery	100% All costs expensed for cost recovery purposes
Profit Oil Split	Negotiable—Published Minimum Terms:
Sliding Scale	**MBOPD** **Gvt./Contr.**
	0–20 30/70%
	20–30 35/65
	30–40 40/60
	40–50 45/55
	50–60 50/50
	60–70 55/45
	70–80 60/40
	80–90 65/35
	>90 70/30
Profit Gas Split	70/30% in favor of the Contractor
Taxation	30% Income tax
	TLCF 4 years
Depreciation	100% Exploration & development for C/R
	100% Exploration costs
	20% SLD
Ringfencing	Yes
Gvt. Participation	None (Although Hydrocarbons Law, Decree 109-83
	stipulated minimum 30%)

Government Take				Effective Royalty Rate	Lifting	Savings Index	Data Quality
Downside	**Mid-range**	**Upside**	**Margin**				
74%	67%	64%	62%	20%	64%	45¢	Fair

INDIA

Late 1980s Various Negotiated PSCs

Area	Not fixed 4th Round 80,000–93,000 km²
Duration	**Exploration** 3 + 2 + 2
	Production 25 + 5 including exploration
	+ 10 for non-associated gas
Relinquishment	25% and 50%
	1986 terms 25% each phase
Exploration Obligations	Negotiable
Bonuses	None
Royalty	None
Cost Recovery	100% no limit—may bid a lower limit, All costs expensed
Profit Oil Split	Investment Multiple (Slightly similar to an R Factor)
	Cumulative Net Cash Flow / Exploration & Development Costs

Investment Multiple	BHP 1988
0 to 1.0	95%
1.0 to 1.5	95
1.5 to 2.0	92.5
2.0 to 2.5	70
2.5 to 3.0	55
3.0 to 3.5	30
3.5 and up	30

Taxation	50%
Depreciation	50% Exploration "When placed in service"
	50% Development "When spent"
Ringfencing	Development costs ringfenced by field.
	Exploration costs are not
Gvt. Participation	40% [10% Heads up + 30% Back in]
DMO	100% at market price

Government Take				Effective Royalty Rate	Lifting	Savings Index	Data Quality
Downside	Mid-range	Upside	Margin				
72%	73%	77%	83%	0%	92.5%	45¢	Good

INDIA

Command, Videocon, Marubeni, ONGC Ravva PSC 28 October, 1994

Area	Ravva Field 331.21 km² (81,809 acres)
Duration	25 years from "Effective Date" 28 October, 1994
Relinquishment	Voluntary
Minimum Work Commitment	$218 MM Dev. $43 MM Exp.
	$33 MM ONGC Carry + through 3-Month "Transfer Period"
Bonuses **Signature**	$ 6.25 MM + 6.25 MM a year later
Production	$ 9.0 MM at 25, 50 and 75 MMBBLS cumulative
	$ 1.8 MM at 80, 85, 95 and 100 MMBBLS cumulative
	Gvt. of India additional crude entitlement:
	Year 3—200,000 BBLS, Year 4 150,000 BBLS
	Year 5—100,000 BBLS, Year 6 50,000 BBLS
Royalties *	Rs. 481 per tonne ≈ $1.80/BBL "Royalty" **
	"Specific Rate" Rs. 900 per tonne ≈ $3.38/BBL "Cess" **
	For Gas 10% Royalty and no "Cess"
	** Assuming Rs. 36/$ and 7.4 BBLS/Tonne
Cost Recovery	Cannot exceed Base Development Costs by > 5%
	Recovery of Past Costs shall not exceed $55 MM
Depreciation	100% for cost recovery purposes; 25% DB for tax
	Abandonment Accrued UOP fund - cost recoverable

Profit Oil Split	**PTRR**	**Contractor/Gvt.**
(Pre-Tax)	0–15%	90/10%
Based on Post	15–20	85/15
Tax ROR method	20–25	80/20
(PTRR)	25–30	75/25
	30–40	65/35
	>40	40/60

Income Tax	50%
Ringfencing	Yes
DMO	Up to 100% of Entitlement @ market price
Gvt. Participation	ONGC 40% JV Partner

Government Take				Effective Royalty Rate	Lifting	Savings Index	Data Quality	ERR depends strongly on price of oil and Rupie/$ exchange rate.
Downside	**Mid-range**	**Upside**	**Margin**					
100%	88%	86%	84%	28+%	70%	44¢	Good	

INDONESIA

Offshore Northwest Java "The First PSC" 18 August, 1966

Area	14,000,000 acres Offshore Sunda Basin
Duration	**Exploration** 6 years with 4-year extension
	Term 30 years from "effective date"
Relinquishment	If no discovery within 10 years, contract terminates
	25% of original area after 3 years and 6 years
Exploration Obligations	$7.5 MM over 6 years
	1st, 2nd, 3rd, 4th, 5th & 6th years
	$300K, $100K, $1,200K, $1,600K, $1,800K & $2MM respectively
Royalty	Nil
Signature Bonus	
Cost Recovery	40% of gross production
Depreciation	10%/year
Profit Oil Split	65/35% (in favor of Government)
Profit Gas Split	Not specified in original contract
Taxation	None [included in Permina (Gvt.) share]
Ringfencing	Each License Ringfenced
DMO	After 60 months production from a field Contractor receives $0.20
	per barrel for 25% of "share oil"
Gvt. Participation	0%

Government Take				Effective Royalty Rate	Lifting	Savings Index	Data Quality
Downside	**Mid-range**	**Upside**	**Margin**				
87%	68%	68%	68%	39%	58%	35¢	Good

INDONESIA

Standard Pre 1984

Area	No restriction—Designated Blocks
Duration	**Exploration** 4 years
	Production 30 years
Relinquishment	25% or 100% of no discovery
Exploration Obligations	Multi-well commitments
Signature Bonus	Various
Production Bonus	Many variations each contract is different
Royalty	Nil
Cost Recovery	No limit
	20% Investment Credit applies to facility, platform, pipeline costs is recoverable but taxable
Depreciation	Oil 7 Year DDB going to straight line in year 5
	Gas 7 Year DB switching to straight line in year 8
Profit Oil Split	65.9091/34.0909% (In favor of Government)
Profit Gas Split	20.4545/79.5455% (In favor of Contractor)
Taxation	56% Effective Tax Rate
	Based on 45% Corporate Income Tax + 20% W/H (Dividend) tax
Ringfencing	Each License Ringfenced
DMO	After 60 months, production from a field Contractor receives $0.20 per barrel for 25% of "share oil"; "Share Oil" is equal to 79.5455% of Contractor entitlement (cost oil and profit oil).
Gvt. Participation	10% Local Company—Option seldom exercised (assumed zero)

Government Take				Effective Royalty Rate	Lifting	Savings Index	Data Quality	Commerciality terms limited contractor entitlement to 49%.
Downside	Mid-range	Upside	Margin					
88%	87%	87%	87%	0%	49%	15¢	Good	

The First PSCs Indonesia

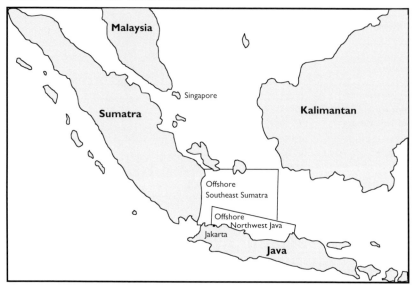

	Northwest Java	Southeast Sumatra
Contractor	IIAPCO	IIAPCO
Contract Signing	18 August, 1966	6 September, 1968
Original License Area	14,000,000 Acres	32,000,000 Acres
Signature Bonus	$1 MM (?)	$1.25 MM + $1.25 MM 6 months later
Work Program (Firm)	$2.1MM (1st3 years)	$22 MM (1st10 years)
1992 Status (at time of contract renegotiations/renewal)		
Production	120,000 BOPD	160,000 BOPD
Remaining (Proved) Reserves	405 MMBBLS	477 MMBBLS
# Fields	34 Old, 14 New	22 Fields
# Platforms (structures)	180	46 (41 Wellhead + 5 Processing)
Producing Wells (beginning of year)	426 wells	224 wells
Average production rate/well	280 BOPD/well	700 BOPD/well

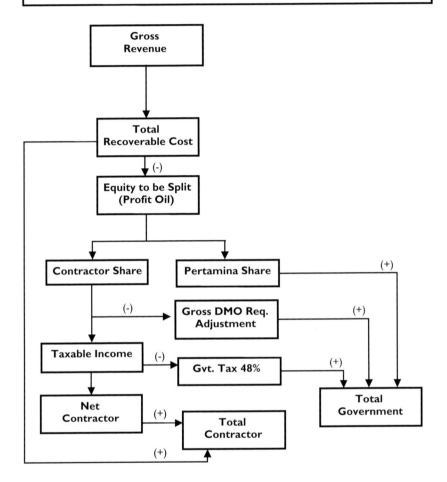

Indonesian Standard PSC

From: **BLOK DIAGRAM PERHITUNGAN KEEKINOMIAN**
Gatot K. Wiroyudo, Chairman, PSC Management and Supervisory Body,
Pertamina, January, 2000 Langkawi Island, Malaysia

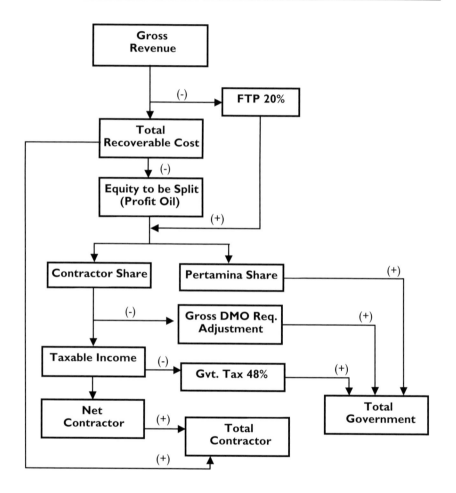

Indonesian Standard PSC with FTP

Gross Revenue

(-) → FTP 20%

Total Recoverable Cost

(-)

Equity to be Split (Profit Oil)

(+)

Contractor Share

Pertamina Share

(-) → Gross DMO Req. Adjustment

Taxable Income

(-) → Gvt. Tax 48%

Net Contractor

(+) → Total Contractor

Total Government

(+)

From: **BLOK DIAGRAM PERHITUNGAN KEEKINOMIAN**
Gatot K. Wiroyudo, Chairman, PSC Management and Supervisory Body, Pertamina, January, 2000 Langkawi Island, Malaysia

INDONESIA

Third Generation (Generasi III) – Post 1988/89

Area	No restriction—Designated Blocks
Duration	**Exploration** 3 years
	Production 20 years
Relinquishment	25%
	or 100% of no discovery
Exploration Obligations	Multi-well commitments
Signature Bonus	Still exist—various (Bonus is not cost recoverable,
	but it is tax deductible)
Production Bonus	Many variations each contract is different
Royalty	Nil
Cost Recovery	80% limit (effectively) because of 1st Tranche Petroleum of 20%
	17% Investment Credit applies to facility, platform, pipeline costs
	Investment Credit is cost recoverable but not tax deductible
Depreciation	Oil 25% declining balance with balance written off in year 5
For C/R and Tax	Gas 10% declining balance with balance written off in year 5
	Depends upon Grouping I, II, III, 50%, 25% and 10% respectively
Profit Oil Split	71.1574/28.8462% (in favor of Government)
Profit Gas Split	42.3077/57.6923% (in favor of Contractor)
Taxation	48% Effective Tax Rate
	Based on 35% Corporate Income Tax + 20% W/H (Dividend) tax
Ringfencing	Each License Ringfenced
DMO	After 60 months production from a field Contractor receives 10% of
	market price for 25% of "share oil". Share oil is equal to 28.8462%
	of Contractor entitlement (cost oil and profit oil).
Gvt. Participation	10% Local Company—Option seldom exercised

Government Take				Effective Royalty Rate	Lifting	Savings Index	Data Quality
Downside	**Mid-range**	**Upside**	**Margin**				
87%	87%	87%	86%	14.2%	48.7%	15¢	Good

IRAN

NIOC/Bow Valley/Bakrie Interinvestindo "Buy-back" - July, 1997

Area	Balal Field
Duration	Until full cost recovery or maximum of 5 yrs from 1st production
Obligations	"Master Development Plan" $169 MM Ceiling
	Any change in expenditures > $25,000 can constitute "Changes in Scope" that can increase the overall $169 MM ceiling.
Cost Recovery	Out of up to 65% of Gross Production
	If full recovery has not occurred after 3 years an additional 2 years is permitted.
	May recover Petroleum Costs and bank finance charges (Interest cost recovery at a rate of 0.75% + LIBOR [compounded monthly] estimated at around $33 MM)
	Petroleum Costs incurred prior to "First Production" and remuneration fee recovered as follows:
	1st Year 40% + Bank Charges
	2nd Year 30% + Bank Charges
	3rd Year 30% + Bank Charges
	Cost incurred after to "First Production" amortized over remainder of the "Amortization Period."
Remuneration Fee	$78.585 MM Remuneration Fee
	[Fees paid in Lavan (Island) Blend crude unless a particular payment is less than a normal cargo size then payment is in US$].
	This is a fixed fee unless JMC approves an adjustment of greater than 10% to $169 MM MDP [Through Changes in Scope] then fee is adjusted proportionately.
Procurement	At least 30% of Petroleum Costs must be Iranian Content
	> $50,000 goods - approved vendors list (of at least 4)
	> $100,000 services - approved subcontractors list
	$500,000 NIOC approval if consistent with MDP
	(Work program and budget)
	< $500,000 NIOC approval if not consistent with MDP
Gvt. Participation	Joint Management Committee 3+3
	Title to facilities and operatorship transfers to NIOC upon commissioning
Other	Training = 2% of CAPEX/Year Cost recoverable as an operating cost (not a capital cost)

IRAN

The Buy-back Contracts

Offshore Fields	Contractor	
Siri A & E	TotalFina, Petronas	The first "buyback" 1995. Sirri A started October, 1998 with 7,000 BOPD. Sirri E started January, 1999 at 30,000 BOPD. Target 100,000 BOPD by 2000.
Balal	Bow Valley, Elf	80 MMBBL field. Target 40,000 BOPD by 2001. This is a production to reserves ratio of 18%!
South Pars	TotalFina, Petronas, Gazprom	2nd and 3rd phases of South Pars Gas field. To provide gas injection onshore to boost production by 300,000 BOPD(?)
Doroud	Elf, ENI	Gas and water flood to increase output from 130,000 to 220,000 BOPD, then up to 290,000 BOPD by 2005(?)
Nowrooz	Shell	70 MMBBL recoverable. Target 90,000 BOPD by 2003. This is a production to reserves ratio of 47%!?
Soroush	Shell	400 MMBBL field. Target 100–150,000 BOPD. Production to reserves ratio of 9–14%
South Pars	ENI, Petro Pars 60/40%	$3.8 Billion, 529 MMCFD + 80,000 B/D NGLs + 1.5 MM tonnes/year LPGs. (150 BBLS NGL/MMCF ?) 4th and 5th phases of South Pars gas for export to Pakistan
Onshore Fields		
Darkhovien	ENI, BP Amoco, Lasmo	700 MMBBLS (with 23% recovery factor) 200,000 BOPD by 2005–2006
Ahwaz-Bangestan		160,000 BOPD by 1998, 500,000 BOPD by 2006
Cheshmeh Kush		Target 80,000 BOPD from 9,000 BOPD
Dehloran		Target 75,000 BOPD for 100 MMBBL field Production to reserves ratio of 27%!?
Mansuri		Target 200,000 BOPD by 2004

Portions from: Fesharaki, F.: Varzi, M.: *Oil & Gas Journal*, 14 Feb., 2000 pp. 44
(Sourced Dresdner Kleinwort Benson)

Estimated Rates of Return (from MEES Email)

Sirri A	20%
Sirri E	23%
South Pars Gas	18%
Balal	24% subject to change

"Alternative Oil" (for reimbursement and remuneration) defined in the Total contract as production from other buy-back contracts only. However, some discussion about provision for expanding this to include proceeds from the sale of so called "trading oil" from Kharg Island exports.

NIOC expects to severely limit this provision except perhaps for North Pars gas where liquids (4 BBLS/MMCF) might not be adequate. South Pars relatively wet @ 40–50 BBLS/MMCF.

Parliament abolished "Alternative Oil" in later contracts and reduced contractor typical maximum share of 65% to 60%. (Bahrain 99)

LIBYA

1990 Model Contract

Duration	
Exploration Obligations	
Royalty	None
Bonuses	None
Cost Recovery Limit	35%
Depreciation	No depreciation for cost recovery

Profit Split
(based on two elements)

BOPD	Production Index *	R Factor	Example R Factor Index ** A Factor
Up to 10,000	.95	0–1.5	1.00
10,000–25,000	.84	1.5–3.0	.80
25,000–50,000	.60	3.0–4.0	.65
50,000–75,000	.30	> 4.0	.50
>75,000	.15		

* Base factor; ** A factor (various rates exist)

For example: Contractor share of "profit" at 25 MBOPD
and R of 1.7 = 70.7%
[.884 (wtd. average at 25 MBOPD) * .8]

Taxation	In lieu (Rate used is 65%)
Ringfencing	Yes
Gvt. Participation	65% Carried through exploration.
	No reimbursement of past exploration costs.
	Contributes 50% of development Costs
	Contributes 65% of operating Costs
	Receives 65% of gross production

Government Take				Effective Royalty Rate	Lifting	Savings Index	Data Quality
Downside	Mid-range	Upside	Margin				
73%	76%	79%	83%	14%	89%	77¢	Fair

Libyan 2-Dimensional Profit Oil Split
1990 Model

Production (BOPD)	Contractor Share of Profit Oil			
> 75	10%	8%	6.5%	5%
50 – 75	20%	16%	13%	10%
25 – 50	50%	40%	32.5%	25%
10 – 25	80%	64%	52%	40%
0 – 10	95%	76%	61.75%	47.5%
A Factor	1.0	.8	.65	.5
R Factor	< 1.5	1.5 - 3	3 - 4	> 4

MALAYSIA

PSC 1994

Area	No restriction—Designated Blocks	
Duration	**Exploration**	3 years + 2 year extension
	Development	2 years + 2 year extension
	Production	15 years for Oil/20 years for Gas
Relinquishment	No interim relinquishment	
Exploration Obligations	Seismic and multi-well commitments	
Bonuses	None (Older contracts had signature and production bonuses)	
Royalty	10% + 0.5% Research Cess	
Cost Recovery	50% limit for Oil/60% for Gas	
	All costs expensed	
Profit Oil Split	**BOPD**	**Split**
(In favor of Government)	Up to 10,000	50/50%
	10,001–20,000	60/40
	> 20,001	70/30
	All production in excess of 50 million barrels 70/30%	
Profit Gas Split	For first 2.1 TCF	50/50%
(In favor of Government)	After 2.1 TCF produced	70/30%
Taxation	40% Petroleum Income Tax	
	20% Duty on Profit Oil Exported (with 50% Export Tax Exemption)	
Depreciation	10 year SLD	
Ringfencing	Each License Ringfenced	
	Gas development costs recovered from Gas production, Oil costs recovered from oil production.	
DMO	None	
Gvt. Participation	Up to 15% Carried through all expl. expenditures	
Other	Supplementary payment of 70% of value of crude oil above the base price.	

National Depletion Policy (NDP) – calls for a 5-year delay of the development of newly discovered fields and a maximum annual rate of depletion amounting to 3% (*The Star*, Thursday December 8, 1994—some reports say 4%) of oil initially in place for fields larger than 400 MMB-BLS oil in place. (First found in 1976 PSCs?).

Government Take				Effective Royalty Rate	Lifting	Savings Index	Data Quality
Downside	**Mid-range**	**Upside**	**Margin**				
90%	83%	84%	86%	32%	54%	22¢	Good

MALAYSIA

Deepwater Terms 1994

Area	No restriction—Designated Blocks
Bonuses	None
Royalty	10%
	0.5% Research Cess
Cost Recovery	75%
	All costs expensed

Profit Oil Split	**BOPD**	**Split**
(In favor of Government)	Up to 50,000	86/14%
	50,001–100,000	82/18
	100,001–200,000	75/25
	> 200,000	63/37
	All production in excess of 300 million barrels 50/50%	

Taxation	40% Petroleum Income Tax
	20% Duty on Profit Oil Exported (with 50% Export Tax Exemption)
Depreciation	20 year SLD
Ringfencing	Each License Ringfenced
	Gas development costs recovered from Gas Production, Oil development costs recovered from oil production.
DMO	None
Gvt. Participation	Up to 15% Carried through exploration with no reimbursement.
Other	Supplementary payment of 70% of crude oil value above base price of around $25.00/BBL + 5% per year

Government Take				Effective Royalty Rate	Lifting	Savings Index	Data Quality
Downside	Mid-range	Upside	Margin				
71%	67%	66%	65%	13%	81%	46¢	Good

MALAYSIA

R/C PSC Model 1997±

Duration	29 years from effective date; Exploration 5 years
	Production 20 years for Oil or expiry of the contract
	20 years + 5 year holding period for Gas
Relinquishment	No interim relinquishment
Exploration Obligations	Bid items
Bonuses	None
Royalty	10% + 0.5% Research Cess

Profit Oil Split and Cost Recovery

Individual Field Total Hydrocarbon Volume (THV) = 30 MMBBLS or 0.75 TCF

Contractor's R/C Ratio	Cost Oil (Gas) Limit	Petronas Share Profit Oil (and Gas)			
		Cumulative Production Below THV		Cumulative Production Above THV	
		Unutilized C/O Split	Normal P/O Split	Unutilized C/O Split	Normal P/O Split
0–1.0	70%	N/A	20%	N/A	60%
1.0–1.4	60%	20%	30%	60%	70%
1.4–2.0	50%	30%	40%	60%	70%
2.0–2.5	30%	40%	50%	60%	70%
2.5–3.0	30%	50%	60%	60%	70%
> 3.0	30%	60%	70%	80%	90%

Price Cap Formula	70% of value of Contractor P/O or P/G above Base Price paid to Petronas. Base price is US$25.00/BBL or $1.80/MMBTU increased by 4% commencing on the 1st anniversary of the Effective Date. But the Price Cap Formula only kicks in if the R/C > 1.0.
Taxation	40% Petroleum Income Tax
(Assumed)	20% Duty on Profit Oil Exported (with 50% Export Tax Exemption)
Depreciation	(?)
Ringfencing	Each License Ringfenced
DMO	None
Gvt. Participation	Up to 15% Petronas carried through all expl. expenditures (Assumed)

Government Take				Effective Royalty Rate	Lifting	Savings Index	Data Quality
Downside	Mid-range	Upside	Margin				
79%	82%	84%	92%	18%	54%	20¢	Good

MALAYSIAN

New R/C Contract Features—1997
(Shell Sarawak is said to have signed the first R/C Contract)

---

(1) R/C Index - for Cost Recovery Limit and Profit Oil Split (Modified *R* Factor)

R/C = Cumulative (Profit Oil Share + Cost Recovery)/Cumulative Contractor Costs

Where cost = Exploration + Development + Operating Costs

It was proposed that contractor would have a one-time option to ringfence the R/C Index at field or contract level.

---

(2) Cost Recovery Tranche

Percentage of gross production available for cost recovery would be on a sliding scale based on the R/C Index.

Unused Cost Oil [Ullage] subject to a different split with the greater share going to the Contractor.

---

(3) Price Cap Formula

Only in effect if R/C > 1

Malaysian officials at Asian Oil and Gas Conference in KL 29 May, 2000 discussed the new "R/C Incentive" contracts pointing out that 14 such contracts had been signed. Their explanation of Government Take is summarized below:

Total Costs	26%	(18% + 8% — presumably OPEX + CAPEX)
Government share	40%	
Petronas share	14%	
Contractor share	20%	(This indicates a Government Take of 73%)
Total Revenues	100%	

However, the speaker quoted Government Take as 54%, thus effectively focusing on division of revenues [40 + 14% = 54%] as opposed to division of profits [(40 + 14%)/(100 − 26%) = 73%].

NIGERIA

PSC 1994 Offshore (Shell and Elf)

Area	Two Offshore Blocks 1994
Duration	**Exploration** 10 yrs (work to be completed in first 6 years)
	Production 20 yrs (for total of 30 yrs)
Relinquishment	50% no later than 10 years from effective date
Exploration Obligations	$176 MM over 10 years
	[$114 MM Elf Nigeria Contract - 5 blocks]
Signature Bonuses	$30 Million for each block for Shell (assumed for Elf)
Production Bonuses	0.2% of 50 MMBBLS cumulative production (from model)
	0.1% of 100 MMBBLS cumulative production at the price
	on the due date.
Rentals	3 Naira/ km² expl; 750 Naira/ km² prod
	[80 Niara/US$1 - Nov., 1996]
Royalty	**Water Depth**
	(meters) **Rate**
	500–800 8%
	800–1000 4
Taxation	50% Petroleum Profits Tax
	50% Investment Credit on Development capital
	excluding intangible drilling costs
Cost Recovery Limit	100%
Depreciation	Exploration costs expensed, Development costs 5 year SLD
Profit Oil Split	**MBOPD** **Split %**
(In favor of Government)	Up to 10 50/50
	10–20 60/40
	Elf contract based on cumulative production (1993—5 blocks)
	Gvt. 20% < 200 MMBBLS; 45% for 200-1,000 MMBBLS;
	60% for 1,000–2,000 MMBBLS.
Ringfencing	Yes
DMO	Gvt. may have contractor purchase NNPC entitlement at discount.
Gvt. Participation	None

	Government Take				Effective Royalty Rate	Lifting	Savings Index	Data Quality
Example	Downside	Mid-range	Upside	Margin				
Shell	75%	77%	77.5%	78%	6%	49%	20¢	Good
Elf	66%	64%	63%	62.5%	6%	58%	40¢	Fair

OMAN

Conquest 1989

Area	1,390 km²
Duration	**Exploration** 2 + 2 + 2 years
	Production 30 years + 10 yr extension
Relinquishment	25% at end of 6th yr + 50% (original) 10th yr
Obligations	1st Reprocess 850, acquire 300 km seismic — $2MM
	2nd & 3rd phases 1 well, $5MM — each
Bonuses	Signature $.20 MM
	Discovery $.25 MM ≈
	Production $1 MM @ 25, 50, 75, 100 and 125, etc MBOPD
Rental	$.35M/year
Royalty	None
Cost Recovery Ceiling	50% for oil < $17/BBL
	40% for oil between $17 and $21/BBL
	30% for oil > $21/BBL
	60% for gas production
Depreciation	None
Profit Oil Split	80/20% in favor of Government
Profit Gas Split	70/30% in favor of Government
Taxation	Taxes paid by National Oil Co on behalf of contractor (55%).
Ringfencing	Yes
DMO	None
Gvt. Participation	None

Government Take				Effective Royalty Rate	Lifting	Savings Index	Data Quality
Downside	**Mid-range**	**Upside**	**Margin**				
90%	80%	80%	80%	40%	44%	20¢	Good

PAKISTAN

Onshore Royalty/Tax System
Zones I, II, III — 1994 Petroleum Policy — 26 Blocks awarded in 1994

Area	3 Zones	1) High Risk High Cost
		2) Med Risk High Cost
		3) Med Risk High to Low Cost
Duration	3 + 5 + 1 year extensions	
Relinquishment	30% + 20%	
Exploration Obligations		
Bonuses	$0.5 MM at start up;	
	$1.0 MM at 30 MM BBLS	
	$1.5 MM at 60 MM BBLS	
	$3.0 MM at 80 MM BBLS	
	$5.0 MM at 100 MM BBLS	
Social welfare fund	$20K/year up to 250K at 50 MBOPD	
Annual Training Fee	$100K pre and $250K post production	
Offshore bonus	2% of approved R&D projects after discovery	
Royalty	12.5%	
Taxation	50.0% Zone 1 Corporate Income Tax	
	52.5% Zone 2	
	55.0% Zone 3	
	15.0% Withholding Tax	
Depletion Allowance	lesser of 15% Gross or 50% of taxable income	
Depreciation	10% Declining Balance	
	Below ground drilling costs expensed	
Ringfencing	Not for dry-hole costs	
	Yes for other costs	
Gvt. Participation	5% Initial increasing upon discovery to:	
	15% Zone 1 **	
	20% Zone 2 **	
	25% Zone 3 **	** no reimbursement of past costs
Other	Import duties 3%	

Example	Government Take				Effective Royalty Rate	Lifting	Savings Index	Data Quality
	Downside	Mid-range	Upside	Margin				
Zone 1	62%	56%	56%	57%	12.5%	87.5%	50¢	Good
Zone 2	65%	60%	60%	61%	12.5%	87.5%	47.5¢	
Zone 3	70%	64%	64%	65%	12.5%	87.5%	45¢	

PAKISTAN

Offshore PSC   1994

Bonuses	$1.0 MM at start up;
	$2.0 MM at 33 MM BBLS
	$5.0 MM at 100 MM BBLS
Social welfare fund	$20K/year up to 250K at 50 MBOPD
Annual Training Fee	$20K pre and $100K post production
Offshore bonus	2% of approved R&D projects after discovery
Royalty	0% 1st 4 years of production
	5% 5th year
	10% 6th year
	12.5% thereafter
	Royalty creditable for income taxes (?)
Cost Recovery Limit	85%

Profit Oil Split   **Profit Gas Split**

Cumulative Production (MMBBLS)	Gvt./Contr. Split %	Cumulative Production (BCF)	Gvt./Contr. Split %
Up to 100	25/75	Up to 600	15/85
100–200	30/70	600–1,200	20/80
200–400	35/65	1,200–2,400	25/75
400–800	50/50	2,400–4,800	40/60
800–1,200	70/30	4,800–7,200	60/40
> 1,200	80/20	> 7,200	75/25

Taxation	40% Corporate Income Tax
	[Possible to be direct or in lieu—assumed here to be direct]
	15% Withholding Tax
Depletion Allowance	Lesser of 15% of gross revenues or 50% of taxable income
Depreciation	20% DB Development costs
	25% SLD Exploration and intangible costs
Ringfencing	Not for dry-hole costs; Yes for other costs
Gvt. Participation	None
Other	Import duties 3%

Government Take				Effective Royalty Rate	Lifting	Savings Index	Data Quality
Downside	**Mid-range**	**Upside**	**Margin**				
56%	55%	56%	57%	4%	82%	38¢	Good

PERU

The term *Peruvian Model* is sometimes used in reference to PSCs. By definition under the Peruvian Type PSC gross production is divided, not *profit oil* as under the classic Indonesian Type PSC. However, the same model could be applied to other systems. In fact in Peru, there was a time when a company could choose any type of system it wanted, PSC, service agreement, R/T system. These all could be considered to be consistent with the Peruvian Model or Peruvian type of system, as shown below. Countries that use this arrangement include Indonesia (JOA/JOB contracts), Peru, Honduras, Tunisia, Algeria, Mexico (Pemex – Government) and Gabon. Trinidad & Tobago had a similar contract in the mid-1970s.

Peruvian PSC

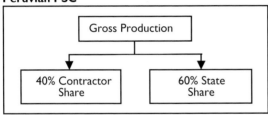

Peruvian Risk Service Agreement

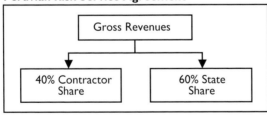

Peruvian Royalty Tax System

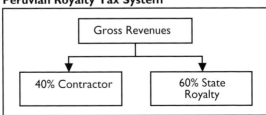

PERU

"License" Contracts—Concessionary 1993 Law/Dec. 1994

Area	Various 1,500 to 10,000 km² blocks
Duration	**Exploration** 7 yrs max with sub-periods typically 2 + 5 1-yr extensions
	Production 30 yrs Oil, 40 Years Gas
Relinquishment	Negotiated—not required in model contract
Exploration Obligations	Various—$20MM–$60MM over 7 years
Bonuses	No Bonuses or Rentals
	Annual training fees from $25K to $50K during exploration; from $50K to $150K during production linked to production levels
Royalty	Negotiated—based on *R* factor typically or production levels

	1994	< $15/BBL			> $15/BBL		
R	**Model**	**Min**	**Typical**	**Max**	**Min**	**Typical**	**Max**
0–1	15%	19	19	45	23	23	51
1–1.5	20%	22	25	49	27	29	52
1.5–2	25%	27	30	56	33	34	56
> 2	35%	35	38	60	37	40	60

Taxation	30% Income Tax (down from 35% in 1991), TLCF limited to 4 years
	0.5% business equity tax (net assets)
	5% labor participation on pre-tax income around all E&P operations (introduced in 1994)
	0% Withholding tax from 1994 (was 7.5% on subcontractors)
Depreciation	5 year SLD or UOP
Ringfencing	No—around upstream activities
Gvt. Participation	None—previously was 25% minimum

Government Take				Effective Royalty Rate	Lifting	Savings Index	Data Quality
Downside	**Mid-range**	**Upside**	**Margin**				
68%	63%	63%	61%	23%	70%	65¢	Good

PERU

Murphy Oil Contract (1995 ?)   Royalty/Tax

Area	Lot 71 1,259,236 Has.
Duration	**Exploration** 6 Periods 18, 12, 12, 18, 12, 12 months
	Production Gas "Holding Period" - 10 Years
	At end of 4th year of exploitation relinquish all but 5 km "halo"
Relinquishment	20% and 20% after 1st and 2nd exploration periods
Exploration Obligations	Reprocess/Reinterpret 950 km seismic 1st period (minimum)
	starting in 3rd Period 1,2,1,1 Wells respectively
Bonuses	

Royalty
(Cash)

R Factor	< $15	$25	>$35
0–1	18.75%	22.5%	28.75%
1–1.5	24	28	34.5
1.5–2	29	35	40
2 +	38.5	42.25	47

$R = X/Y; X = Accumulated\ Revenues, Y = Accumulated\ Disbursements$

Cost Recovery	None
Taxation	30% down from 35% in 1991
	1% withholding tax
	5% Other
Depreciation	5 year SLD
Ringfencing	No
Gvt. Participation	No

Government Take				Effective Royalty Rate	Lifting	Savings Index	Data Quality
Downside	Mid-range	Upside	Margin				
69%	63%	65%	66%	22.5%	70%	65¢	Good

The PHILIPPINES

Risk Service Contract early 1990s

Area	Designated Blocks
Duration	**Seismic Option** 1 yr
	Exploration 10 yrs maximum
	Production 30 yrs
Exploration Obligations	Negotiable; Two Well Option after Seismic
Signature Bonus	Negotiable
Royalty	None
FPIA	7.5% goes to contractor group
	*Filipino Participation Incentive Allowance (FPIA)
	Depends on level of Filipino ownership up to 30% onshore or 15%
	in deepwater qualifies for full 7.5% (FPIA)

Onshore Filipino Participation	FPIA
Up to 15%	0%
15–17.5	1.5
17.5–20	2.5
20–22.5	3.5
22.5–25	4.5
25–27.5	5.5
27.5–30	6.5
> 30	7.5%

Cost Recovery	70% limit
Depreciation	Exploration costs expensed, Development costs 10 year SLD
Profit Oil Split	60/40% (In favor of Gvt.)
	[Contractors 40% is part of "Service Fee"]
Taxation	No - Paid out of Gvt. Share
Ringfencing	Consolidation allowed on 2 or more deepwater blocks
DMO	Pro-rata
Gvt. Participation	None

Government Take				Effective Royalty Rate	Lifting	Savings Index	Data Quality
Downside	Mid-range	Upside	Margin				
50%	53%	54.5%	55.5%	13.5%	62.5%	40¢	Good

TRINIDAD & TOBAGO

BHP/Elf Offshore PSC 29 Feb., 96

Area	Block 2(c) 51,766 hectares
Duration	**Exploration** 3+2+1 + development phase Gas discoveries **Production**
Relinquishment	30 + 20% on commercial discovery retain up to 20%
Bonuses Production	$1 MM @ 25, $1.5 MM @ 50, $2 MM @ 75, $2.5 MM @ 100 MBOPD + $1 MM for every 50 MBOPD >100 MBOPD
Training	$100K + 6%/yr Expl., $150K + 6%/yr after discovery
R&D fee	$100K + 6%/yr Expl., $150K + 6%/yr after discovery
Technical Assistance Bonus	$100K
Administrative fee	$200K + 6%/yr
Annual rentals	$1.25/hectare 1st year, $1.50, $1.75, $2.00, $2.25, $2.50 then + 6%/year
Exploration Obligations	1st Phase 20,000 line km 3-D seismic, full fold 642 km², + 1 well (3,500 m) $6.5MM * 2nd Phase 1 well (2,500 m) $3MM * 3rd Phase 1 well (2,500 m) $3MM * * *Bank guarantees required*
Cost Oil	35% Oil 50% Gas
Depreciation	Exploration 100%; Development 20% SLD
G&A	pre discovery 5% of 1st $5MM, 3% next $3MM, 1% over $8MM post discovery .6% up to $20MM, .5% $20-50MM, .4% over $70 MM

Ministers Share Profit Oil/Gas

	Oil Price $/BBL					Gas Price $/MCF			
MBOPD	**<$20**	**20–30**	**30–40**	**>$40**	**MMCFD**	**<$1**	**1–1.5**	**1.5–2**	**>$2**
0–10	50%	50%	55%	60%	0– 60	50%	50%	55%	50%
10–25	50	55	60	65	60–150	50	50	50	50
25–50	55	60	65	70	150–300	55	55	55	55
50–75	60	65	70	75	300–450	60	60	60	60
> 75	65	70	75	80	> 450	65	65	65	65

Taxation	In lieu
Ringfencing	Yes
Gvt. Participation	No (However some contracts have up to 25% Participation)

Government Take				Effective Royalty Rate	Lifting	Savings Index	Data Quality
Downside	**Mid-range**	**Upside**	**Margin**				
77%	52%	53%	55%	33%	63%	46¢	Good

QATAR

PSC 1994 Model

Duration	
Relinquishment	40% after 6 years
Bonuses	Various
Royalty	None
Cost Recovery Limit	40% [30–40%]
	It is assumed that excess cost oil goes to Government
Depreciation	All costs 4 yr SLD

Production Sharing
(Negotiable)
Example Company Share

Profit Oil Share
R Factor

MBOPD	0–1	1–1.5	1.5–2	> 2
0–15	45%	40%	35%	30%
15–30	40	35	30	25
30–45	35	30	25	20
45–60	30	25	20	15
>60	25	20	15	12

Profit Gas Share
R Factor

MMCFD	0–1	1–1.5	1.5–2	> 2
0–130	45%	40%	35%	30%
131–260	40	35	30	25
261–390	35	30	25	20
391–520	30	25	20	15
>520	25	20	15	12

Taxation	Taxes paid out of government share of profit oil
Ringfencing	Yes
Gvt. Participation	None

Government Take				Effective Royalty Rate	Lifting	Savings Index	Data Quality
Downside	**Mid-range**	**Upside**	**Margin**				
63%	70%	78%	86%	34%	49%	27¢	Good

RUSSIA—Sakhalin II

MMMMS Consortium - 23 June, 1994

Area	1,546 km² Piltun-Astikhskoye and Lunskoye fields
Duration	**Exploration** 5 years
	Production 20 years—right to extend
Relinquishment	No mandatory provisions
Obligations	$50 MM Appraisal
Bonuses Signature	0
Other Bonuses	$15 MM @ Commencement Date
	$15 MM @ Development Date Piltun-Astikhskoye
	$20 MM @ Development Date Lunskoye
Rentals	1-2% of work program costs
Payments	$100 MM to Regional Development Fund $20/yr
	beginning with development approval
	$160 MM Reimbursement of Prior Russian Expenditures
	@ $4 MM/Qtr for 20 Qtrs
	Another $4 MM/Qtr for 20 Qtrs
	starting when Russian Profit Oil Split = 70%
Royalty	6%
Cost Recovery Limit	100% after Royalty
	$100 MM Regional Dev. Fund
	+ $160 MM Reimbursement are cost recoverable
Depreciation	All costs expensed (Assumed)

Production Sharing	**Real pre-Tax IRR**	**Government Share**
(Pre-tax)	Less than 17.5%	10%
	17.5% to 24	50
	More than 24	70

Taxation	32% Profit Tax "shall not exceed"
	Fees, interest, bonuses deductible; Tax loss carry forward 15 years
Depreciation	Capital expenditures 3 year SLD (Assumed)
Ringfencing	Yes
Gvt. Participation	None

Government Take				Effective Royalty Rate	Lifting	Savings Index	Data Quality
Downside	Mid-range	Upside	Margin				
48%	69%	72%	81%	6%	62%	34¢	Good

SYRIA

PSC Unocal 23 June, 1992

Area	Block III
Duration	**Exploration** 3+2+1 (I believe)
	Production 20 years typically + 10 yr extension
Relinquishment	25% + 25% of original area each time
Exploration Obligations	
Bonuses	**Signature** $2.5 MM
	Production $1.5, 2, 4, 4, & $8 MM
	at 25, 50, 100, 150, & 200 MBOPD
	Not recoverable
Training	$50,000/year (recoverable)

Royalty	**MBOPD**	**Rate**
	0–50	14%
	50–100	15%
	> 100	16%

Cost Recovery Ceiling	25% up to 50,000 BOPD
"net of royalty"	20% over 50,000 BOPD
	Excess cost oil goes directly to National Oil Corp. - SPC.
Depreciation	Exploration capital and operating costs expensed
	Development Costs 5 years SLD

Profit Oil Split		**Contractor**
	BOPD	**Share**
	Up to 25,000	19.5%
	25,001–50,000	18.5
	50,001–100,000	18.0
	100,001–200,000	15.0
	> 200,000	12.0

Taxation	In lieu "Paid by the State"
Ringfencing	Yes
Gvt. Participation	None

Government Take				Effective Royalty Rate	Lifting	Savings Index	Data Quality
Downside	**Mid-range**	**Upside**	**Margin**				
100%	91%	86%	88%	63%	36%	19¢	Good

TIMOR GAP - ZOCA

Zone A—1991/92 License Round—PSC Jointly Administered by Indonesia/Australia 50/50%

Area	Main blocks in Zone of Cooperation "A" (ZOCA) comprise 20–40 sub blocks at 10 km² each
Duration	Exploration 6 Years with option for 4-year extension With development contract automatically extends to 30 years
Committed Expenditures	Exploration 1st year seismic only $1–4 MM 2nd year 0–2 wells $.5–8 MM 3rd year 1–3 wells $.5–21 MM 4th –6th years 1–4 wells $6–30 MM
Relinquishment	25% after 3 years—another 25% after 6th year
Bonuses	
Royalty	None
Cost Recovery Limit	90% Effective limit for 1st 5 years * 80% Effective limit thereafter * 0% First Tranche Petroleum (FTP) Similar to Indonesian FTP after 5 years production, FTP reverts to 20%
Depreciation	5 year SLD

Profit Oil Split

BOPD	Contractor Share
Up to 50,000	50%
50,001–150,000	40%
150,001–200,000	30%
Natural Gas	50%

Taxation	48% Effective Tax rate (Similar to Indonesia) Comprised of 35% Income Tax and 20% Withholding Tax Companies will lodge income tax returns with both countries. In each country a 50% tax rebate will be given.
Ringfencing	Yes
DMO	Similar to Indonesian DMO 25% of "share oil" (after 5 years, 10% of market price)
Gvt. Participation	None
Other	27% Investment Credit (IC) for deepwater tangible exploration and capital costs for pipeline, platforms, processing facilities (ppp)

Government Take				Effective Royalty Rate	Lifting	Savings Index	Data Quality
Downside	**Mid-range**	**Upside**	**Margin**				
72%	73%	74%	75%	5%	70%	26¢	Good

TURKMENISTAN

Monument PSC 7 Aug., 1996

Area	Nebit-Dag/5 area 1,800 km²		444,600 acres	
Duration	"Effective period" 25 years + optional 10 yr extension			
	Exploration 5 = 2 + 1.5 + 1.5			
	Production 20 + 10 yr extension			
Relinquishment	@ 2 years 30%			
	@ 2 + 1.5 years 20% of remaining			
	@ 2 + 1.5 + 1.5 years 20% of remaining			
	@ 5th year of 3rd (production) period all remaining except production			
Obligations	1st Period	$10 MM min (excess credited)		
	2nd	$40 MM Original oil vs. new oil		
Bonuses				
Royalties	**Based on "New Oil"**	**Royalty**	**Gas MMCFD**	**Royalty**
	Up to 25,000 BOPD	3%	Up to 150	3%
	25,000–50,000	5	150–300	5
	50,000–75,000	7	300–450	7
	75,000–100,000	10	450–600	10
	> 100,000	15	> 600	15
Cost Recovery Ceiling	60% of "Net Production" All costs expensed (Assumed)			
Production Sharing	**"S" Multiplier = P/O Split**		**P**	**S**
	a = contractor cost oil		0–1	40/60%
$P = \dfrac{\sum (a+b)}{\sum (c+d)}$	b = contractor profit oil		1–1.5	50/50
	c = Capex + Libor		1.5–2	60/40
	d = Opex		2–2.5	80/20
			> 2.5	90/10
Taxation	25%			
Depreciation	Capital expenditures 5 year SLD			
Ringfencing	Not for profit tax - for cost recovery yes			
Gvt. Participation	None			

Government Take				Effective Royalty Rate	Lifting	Savings Index	Data Quality
Downside	Mid-range	Upside	Margin				
62%	67%	77%	91%	18%	61%	35¢	Good

VENEZUELA

1996 Risk Service Agreements "Strategic Associations" Round 3 Exploration

Area	8 to 12 areas in blocks not more than 2,000 km² divided into 16 sub-blocks.
Duration	Exploration up to 9 years Total 20 years with option to extend by 10 years
Relinquishment	
Exploration Obligations	2 wells per 1,000 km² in first 4 years If exploration continues into yrs 5–7, 6 more wells required.
Signature Bonuses Data Packages Bid Fee	Initial Guarantee $500,000 $50,000 $100,000 per bid
Royalties	16.67% Proposed sliding scale Based on ROA = Pre-tax profit/Asset Book Value
Taxation	Sliding scale PEG tax levied on pre-tax profits. PEG tax = Extra Government Take; Bid item (0–50%) 67.7% Corporate Income Taxes (interest expense deductible) Investment tax credit limited to 2% of taxable income. VAT 16% [Zero for exports]
Depreciation	Exploration and Development drilling UOP, dry wells expensed
Ringfencing	Yes
DMO	None
Gvt. Participation	Sliding Scale up to 35%

Licences were awarded on the basis of PEG bid (from 0 to 50%). Ties were broken with a bonus to follow within 2 hours.

Government Take				Effective Royalty Rate	Lifting	Savings Index	Data Quality
Downside	Mid-range	Upside	Margin				
93%	91%	88%	87%	16.7%	0%	20¢	Good

ZAMBIA

ROR Based PSC 1986 Model

Area	
Duration	**Exploration** 8 year
	Production 25 years
Relinquishment	negotiated
Exploration Obligations	
Bonuses	None
Rentals	
Royalty	12.5% for Oil in model contract (Minister may specify a lower rate)
	Gas - negotiated case-by-case
Cost Recovery	100% No Limit
Production Sharing	One Example

After-tax	Gvt.
Real ROR	Share
Up to 15%	0%
15–20	50
20–25	60
25 and up	70

Tax	50%
Depreciation	25% per year SLD (Assumed)
Ringfencing	No—Consolidation allowed
DMO	Perhaps if Country is net importer
Gvt. Participation	None assumed

Government Take				Effective Royalty Rate	Lifting	Savings Index	Data Quality
Downside	Mid-range	Upside	Margin				
65%	74%	80%	85%	12.5%	87.5%	50¢	Fair

Glossary

Abrogate – To officially abolish or repeal a treaty or contract through legislative authority or an authoritative act.

Accelerated depreciation – Writing off an asset through depreciation or amortization at a rate that is faster than normal accounting straight-line depreciation. There are a number of methods of accelerated depreciation, but they are usually characterized by higher rates of depreciation in the early years than latter years in the life of the asset. Accelerated depreciation allows for lower tax rates in the early years.

Acreage – Amount of land area (or offshore area) under lease or associated with and/or governed by a production-sharing contract (PSC).

Ad valorem – Latin, meaning *according to value*. A tax on goods or property, based upon value rather than quantity or size. Royalties are typically ad valorem, based upon value *at the wellhead*.

Affiliate – Two companies are affiliated when one owns less than a majority of the voting stock of the other or when they are both subsidiaries of a third parent company. A subsidiary is an affiliate of its parent company. (*see* **Subsidiary**)

Amortization – An accounting convention designed to emulate the cost or expense associated with reduction in value of an intangible asset (*see* **Depreciation**) over a period of time. Amortization is a non-cash expense. Similar to depreciation of tangible capital costs, there are several techniques for amortization of intangible capital costs:

- Straight Line Decline (SLD)
- Double Declining Balance (DDB)
- Declining Balance (DB)
- Sum of Year Digits (SYD)
- Unit-of-Production (UOP)

API – American Petroleum Institute.

API gravity – American Petroleum Institute measure of the density or weight of a crude oil. Measured in degrees (°) as in "West Texas Intermediate is a 38°–40° API crude."

$$°API = \frac{141.5}{Sg} - 131.5$$

Sg = specific gravity in grams per cubic centimeter.

Appraisal well – *See* **Delineation well**.

Aquifer – Porous water-bearing rock.

Arbitration – A process in which parties to a dispute agree to settle their differences by submitting their dispute to an independent individual arbitrator or group such as a *tribunal*. Typically, each side of the dispute chooses an arbitrator, and those two arbitrators choose a third. The third arbitrator acts as the chairman of the tribunal, which then hears and reviews both sides of the dispute. The tribunal then renders a decision that is final and binding.

Backwardation – When a commodity's current prices or spot price is greater than futures prices, the market is said to be *inverted* or in backwardation. (*See* **Contango**, which is the opposite of backwardation).

Basket – This term is often used to mean a hypothetical blend of crudes also referred to as a *coctail* for price reference purposes in the absence of arms-length sales.

Block – A license area or contract area; relates to each individual parcel of acreage held by an oil company or a government.

Book value – (1) The value of the equity of a company. Book value per share is equal to the equity divided by the number of shares of common stock. Fully diluted book value is equal to the equity less any amount that preferred shareholders are entitled to, divided by the number of shares of common stock. (2) Book Value of an asset or group of assets is equal to the initial cost less DD&A (effectively depreciation).

Branch – An extension of a parent company but not a separate independent entity. Subsidiary companies are normally taxed as profits and are distributed as opposed to branch profits, which are taxed as they accrue.

Brown tax – A tax that can be positive or negative. A *cash-flow based* Government (working interest) participation could be viewed this way. During the periods of investment, the government *pays*. During the periods of positive cash flow, the government *takes*.

Bubble point – Reservoir pressure at which gas in solution (in the oil) will bubble out of the host oil at the prevailing reservoir temperature.

Calvo clause – A relatively obsolete contract clause once promoted in Latin American countries where the contractor explicitly renounced the protection of its home government over its operation of the contract. The objectives of the Calvo Doctrine were to direct disputes to local jurisdictions and avoid international arbitration.

Capitalization – All money invested in a company including long-term debt (bonds), equity capital (common and preferred stock), retained earnings and other surplus funds. Market capitalization is stock price times the numbers of shares of common stock.

Capitalization rate – The rate of interest used to convert a series of future payments into a single present value.

Capitalize – (1) In an accounting sense, the periodic expensing of capital costs such as through amortization, depreciation, or depletion. (2) To convert an (anticipated) income stream to a present value by dividing by an interest rate. (3) To record capital outlays as additions to asset value rather than as expenses.

Generally, expenditures that will yield benefits to future operations beyond the accounting period in which they are incurred are capitalized—that is, they are depreciated at either a statutory rate or a rate consistent with the useful life of the asset.

Carried interest – When a working interest partner in the exploration or development phase of a contract is paying a disproportionately lower share of costs and expenses than its working interest share. This occurs when government agencies such as the Oil Ministry or the National Oil Company (NOC) are *carried* through the exploration phase of a contract. In this case, the NOC is said to have a *back-in option*. Also in a farm-in agreement, typically, the company holding the original working interest will farm out a portion of the work obligation to another company and is *carried* through that portion of the work program. The company farming in then *carries* the original license holder through that phase, i.e. the original license holder then does not pay, or pays a disproportionately lower percentage.

Cash flow – Gross revenues less all associated capital and operating costs. Contractor cash flow is equal to gross revenues less all costs, government royalties, taxes, imposts, levies, duties and profit oil share, etc. It therefore represents contractor share of profits. Government cash flow typically consists of government royalties, taxes, imposts, duties, profit oil share, etc.

In a financial sense, net income plus depreciation, depletion and amortization and other non-cash expenses. Usually synonymous with cash earnings and operating cash flow. An analysis of all the changes that affect the cash account during an accounting period. (*See* Fig. G–1)

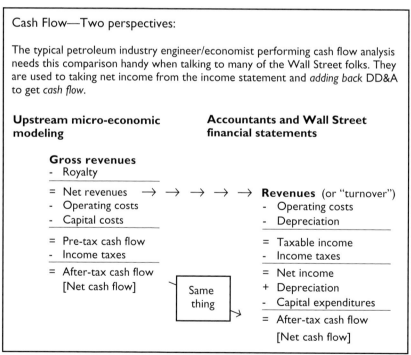

Fig. G–1 *After-tax cash flow*

Central bank – The primary government-owned banking institution of a country. The central bank usually regulates all aspects of foreign exchange in and out of the country. It actively intervenes in the acquisition and sale of its own currency in foreign exchange markets primarily to maintain stability in the value of the country's currency.

Commercial discovery – (Or commercial success) In popular usage, the term applies to any discovery that would be economically feasible to develop under a given fiscal system. As a contractual term, it often applies to the requirement on the part of the contractor to demonstrate to the government that a discovery would be sufficiently profitable to develop from both the contractor and government point of view. A field that satisfied these conditions would then be *granted commercial status*, and the contractor would then have the right to develop the field.

Commingled production – Production of petroleum from more than one reservoir through a single wellbore or flowline without separate measurement.

Completion – Equipment and activities required after drilling a well in order to prepare the well for production of oil and/or gas.

Concession – An agreement between a government and a company that grants the company the right to explore for, develop, produce, transport, and market hydrocarbons or minerals within a fixed area for a specific amount of time. The concession and production and sale of hydrocarbons from the concession are then subject to rentals, royalties, bonuses, and taxes. Under a concessionary agreement, the company would take title to gross production less government royalty oil *at the wellhead*.

Condensate – Light liquid hydrocarbons associated with gas, typically *pentanes plus* (C_5 +). (*see* **Hydrocarbon series**)

Consortium – A group of companies operating jointly, usually in a partnership with one company as operator in a given permit, license, contract area, block, etc.

Contango – The relationship between a commodity's futures prices and the current market price for the commodity. When futures prices are greater than current prices (such as spot prices or current contract prices), the market is said to be in *contango*. Contango is the opposite of *backwardation*.

Contractor – An oil company operating in a country under a PSC or a service contract on behalf of the host government for which it receives either a share of production or fees consisting of reimbursement of costs and remuneration.

Contractor take – Total contractor after-tax share of cash flow.

Cost insurance and freight (CIF) – is included in the contract price for a commodity. The seller fulfills his obligations when he delivers the merchandise to the shipper, pays the freight and insurance to the point of (buyer's) destination and sends the buyer the bill of lading, insurance policy, invoice, and receipt for payment of freight. The following example illustrates the difference between an FOB (free on board) Jakarta price and a CIF Yokohama price for a ton of LNG. (*see* **FOB**).

FOB Jakarta	$170/ton also called "netback price"
	+ 30/ton Freight Charge
CIF Yokohama	$200/ton

Cost of capital – The minimum rate of return on capital required to compensate debt holders and equity investors for bearing risk. Cost of capital is computed by weighting the after-tax cost of debt and equity according to their relative proportions in the corporate capital structure.

Cost oil – A term most commonly applied to PSCs that refers to the oil (or revenues) used to reimburse the contractor for exploration, development and operating costs incurred by the contractor.

Cost recovery – The means by which companies recover costs; same as deductions.

Cost recovery limit – Typically with PSCs in any given accounting period, there is a limit to the amount of deductions that can be taken for cost recovery purposes. The limit is usually quoted in terms of a percentage of gross revenues or gross production. Unrecovered costs are carried forward and recovered in subsequent accounting periods if there is sufficient production.

Country risk – The risks and uncertainties of doing business in a foreign country, including political and commercial risks. (*see* **Sovereign Risk**).

Creeping nationalization or creeping expropriation – A subtle means of expropriation through expanding taxes, restrictive labor legislation, or labor strikes, withholding work permits, import restrictions, price controls, and tariff policies.

Crypto tax – This is a nontechnical reference to non-conventional (less direct) means by which governments may impose duties, levies, or financial requirements on an oil company. These elements rarely are captured in typical published *take* statistics. Examples include: social welfare development funds (written or unwritten), hostile audits, mandatory currency conversions, customs duty exemptions that are not honored, hiring and purchase quotas, inordinately long depreciation rates, inefficient procurement requirements, excessive immigration/visa requirements, etc.

Debt service – Cash required in a given period, usually one year, for payments of interest and current maturities of principal on outstanding debt. In corporate bond issues, the annual interest plus annual sinking fund payments.

Delineation well – A well drilled in order to determine the extent of a reservoir also known as an *appraisal well*.

Depletion – (1) Economic depletion is the reduction in value of a wasting asset by the removal of minerals. (2) Depletion for tax purposes (depletion allowance) deals with the reduction of mineral resources due to removal by production from an oil or gas reservoir or a mineral deposit.

Depletion allowance – This is one type of *incentive* that a few governments use to encourage investment. Typically these allowances provide the companies a *deduction* for tax calculation purposes based on some percentage of gross revenues. The Filipino Participation Incentive Allowance (FPIA) in the Philippines has this characteristic. It allows the contractor group 7.5% of gross revenues as part of the service fee.

Depreciation – An accounting convention designed to emulate the cost or expense associated with reduction in value of a tangible asset due to wear and tear, deterioration or obsolescence over a period of time. Depreciation is a non-cash expense. There are several techniques for depreciation of capital costs:

- Straight-Line Decline
- Double-Declining Balance
- Declining Balance
- Sum-of-Year Digits
- Unit-of-Production

Development costs – Costs associated with placing an oil or gas discovery into production. These costs typically consist of drilling, production facilities, and transportation costs.

Development drilling – Drilling that follows exploratory and appraisal drilling after a discovery.

Development well – A well drilled within a proven or known productive area of an oil or gas reservoir.

Dew point pressure – The (gas) reservoir pressure below which liquids begin to condensate out of the gas at the prevailing reservoir temperature.

Dilution Clause – In a joint operating agreement, a clause that outlines a formula for the dilution of interest of a working interest partner, if that partner defaults on a financial obligation. (*See* **Withering Clause**)

Direct tax - A tax that is levied on corporations or individuals—the opposite of an indirect tax such as a value-added tax (VAT) or sales taxes.

Discounted cash flow analysis – Economic modeling of anticipated income versus expenditures over time. It is based upon estimated production rates, oil prices and costs, as well as royalties, taxes and other means of government take. The net result is a stream of cash flow over time. Cash received in the distant future is not as valuable as cash received now, so the time value of the cash flow is calculated factoring in time value of money to arrive at a *present value* equivalent.

Discounted rate – In discounted cash flow (DCF) analysis, the interest or cost of capital percentage used to calculate time value of money. The present value formula for a *payment* received at some future time n at a discount rate i is:

Present value $=$ future payment$/(1+i)^n$

Where: present value of the future payment is a function of the discount rate i and time n.

For example, if a payment of $5 MM was to be received in 6 years ($n = 6$) the present value discounted at 10% ($i = 0.10$) is.

Present value $=$ $5 MM$/(1+.1)^6 =$ $5 MM$/1.77156 =$ $2.82 MM

Disposal – This term usually refers to transportation and sales of crude or gas from the field.

Dividend withholding tax – A tax levied on dividends or repatriation of profits. Tax treaties normally try to reduce these taxes whether they are so named or simply operate in the same manner as a withholding tax.

Dollars-of-the-day – A term usually associated with cost estimates that indicate the effects of anticipated inflation have been taken into account. For example, if a well costs $5 MM right now in *today's dollars*—(the opposite of *dollars-of-the-day*) then the cost of the well two years from now might be estimated at $5.51 MM in *dollars-of-the-day*, assuming a 5% inflation factor. Other associated terms:

- Dollars-of-the-day   vs.   Nominal Dollars
- Escalated   vs.   Non-escalated
- Current Dollars   vs.   Today's dollars
- Inflated   vs.   Real

Domestic market obligation (DMO) – Some countries provide the state an option to purchase a certain portion of the contractor's share of production. This is called domestic market obligation or domestic requirement. Typically the purchase price for DMO crude is less than market price. Also local currency may be part of the price formula. There are many variations on this theme.

Domestication - A form of creeping nationalization where a host government enacts legislation that forces foreign-owned enterprises to surrender various degrees of ownership and/or control to nationals.

Double taxation – (1) In economics, a situation where income flow is subjected to more than one tier of taxation under the same domestic tax system—such as state/provincial taxes, then federal taxes or federal income taxes and then dividend taxes. (2) International double taxation is where profit is taxed under the system of more than one country. It arises when a taxpayer or taxpaying entity resident (for tax purposes) in one country generates income in another country. It can also occur when a taxpaying entity is resident for tax purposes in more than one country.

Double taxation treaty – Formal agreement between countries to reduce or eliminate double taxation. A bilateral tax treaty is a treaty between two countries to coordinate taxation provisions that would otherwise create double taxation. A multilateral tax treaty involves three or more countries for the same purpose. The United States has few treaties with oil-producing nations.

Dual residence – When a taxpaying entity is resident for tax purposes in more than one country. This can happen when different countries apply the tests for determining residence, and the company passes the test in more than one country.

Dutch disease – The adverse results of large-scale positive shock to a single sector of a nation's economy—so named because of the problems associated with large-scale development of the Groningen Gas field in the Netherlands in the 1970s. Typically the sector of the economy that is booming causes widespread inflation, and other sectors (particularly agriculture) suffer from inability to attract workers. The dramatic increase in foreign exchange can cause problems with local currencies and fiscal, and monetary problems can occur without proper management.

Economic profit – Gross project revenues minus total costs that include exploration, development, and operating costs.

Economic rent – While there are a number of definitions, one common definition is: the difference between the value of production and the cost to extract it. The extraction cost consists of normal exploration, development, and operating costs as well as a share of profits for the industry. Economic rent is what the governments try to extract as efficiently as possible.

Effective royalty rate – The minimum share of revenues (or production) the government will receive in any given accounting period from royalties and its share of profit oil.

Entitlements – The shares of production to which the operating company, the working interest partners, and the government or government agencies are authorized to lift. Entitlements are based on royalties, cost recovery, production sharing, working interest percentages, etc. (*see* **Lifting**)

Excise tax – A tax applied to a specific commodity such as tobacco, coffee, gasoline, or oil; based either on production, sale, or consumption.

Exclusion of areas – (*see* **Relinquishment**)

Expected monetary value (EMV) – (*see* **Expected value**)

Expected value – A weighted average financial value of various possible outcomes such as either a discovery or a dry hole weighted according to the estimated likelihood (estimated probability of success or failure) that either outcome might occur. (*See* **Expected monetary value**)

Expense – (1) In a financial sense, a non-capital cost associated most often with operations or production. (2) In accounting, costs incurred in a given accounting period that are charged against revenues. To *expense* a particular cost is to charge it against income during the accounting period in which it was spent. The opposite would be to *capitalize* the cost and charge it off through some depreciation schedule.

Exploration drilling – Drilling in an un-proved area. (*see* **Exploratory well**)

Exploratory well – A well drilled in an unproved area. This can include: (1) a well in a proved area seeking a new reservoir in a significantly deeper horizon, (2) a well drilled substantially beyond the limits of existing production. Exploratory wells are defined partly by distance from proved production and by degree of risk associated with the drilling. Wildcat wells involve a higher degree of risk than exploratory wells.

Expropriation – Similar to the concept of nationalization or outright seizure or confiscation of foreign assets by a host government. With expropriation, the confiscation is directed toward a particular company; nationalization is where a government confiscates a whole industry. Expropriation is legal but theoretically must be accompanied by prompt adequate and effective compensation and must be in the public interest.

Fair market value (FMV) of reserves – Often defined as a specific fraction of the present value of future net cash flow discounted at a specific discount rate. One common usage defines FMV at two-thirds to three-fourths of the present value of future net cash flow discounted at the prime interest rate plus .75 to 1 percentage point.

Fairway – (*see* **Trend**)

Farmee – The party farming-in.

Farm-in – (1) A lease or working interest obtained from another company in return for a consideration. (2) To receive a farm-in.

Farmor – The party farming-out.

Farmout, farm out – (1) A lease or working interest granted to another company in return for a consideration. (2) To grant a farmout.

Farmout extension – Sometimes the NOC or the government will allow a contractor some additional time (an extension to the current contract phase) to find a partner. Governments know that finding partners is an important way for companies to spread the risk of exploration.

Finding cost – The amount of money spent per unit (barrel of oil or MCF of gas) in exploration divided by reserves added. There are numerous formulas but generally includes discoveries and revisions to previous reserve estimates. Some include acquisition costs of reserves.

Fiscal marksmanship – The ability of authorities to predict with any degree of accuracy or certainty the tax revenues that may fall due to be paid to the government. In the petroleum industry, it is particularly difficult to accurately estimate what revenues may be generated for countries with little or no exploration history.

Also the ability to determine the appropriate taxation scheme

Fiscal system – Technically the legislated taxation structure for a country including royalty payments. In popular usage, the term includes all aspects of contractual and fiscal elements that make up a given government-foreign oil company relationship.

Flare – Or "flaring"; burning of residue hydrocarbon gasses.

Flooding – Injection of water (*water flood*) or gas (*gas flood*) into or adjacent to a reservoir to increase oil recovery.

FOB (Free on Board) – A transportation term that means the invoice price includes transportation charges to a specific destination. Title is usually transferred to the buyer at the FOB point by way of a bill of lading. For example, FOB New York means the buyer must pay all transportation costs from New York to the buyer's receiving point. FOB plus transportation costs equals CIF price. (*See* **Cost Insurance Freight [CIF]**)

FOB shipping point: Buyer bears transportation costs from point of origin.

FOB destination: Supplier bears transportation costs to the destination.

Foreign Corrupt Practices Act (FCPA) – Sometimes referred to as *anti-bribery legislation*. It is illegal for a U.S. company or individual to knowingly pay a bribe to a foreign official in order to obtain or retain business. This includes commissions or payments to agents or intermediaries with the knowledge that all or a part of the payments will be given to a foreign official. The FCPA also has various recordkeeping and reporting requirements.

Foreign tax credit – Taxes paid by a company in a foreign country may sometimes be treated as *taxes paid* in the company's home country. These are creditable against taxes and represent a direct dollar-for-dollar reduction in tax liability. This usually applies to foreign income taxes paid and credited against home country income taxes. Other taxes that may not qualify for a tax credit may nevertheless qualify as deductions for home country income tax calculations.

Formation – A layer of rock or geological horizon that can be mapped. It has a distinct top and bottom. The formation is typically given a name such as the Red Wall Limestone or Kimmeridgian Shale.

Franked dividends – Dividends that have already been taxed at the corporate level and are therefore either not subject to withholding tax or the taxes paid are creditable against withholding taxes.

Gas oil ratio (GOR) – The number of cubic feet of natural gas produced with each barrel of oil produced. It is measured under surface conditions. Also known as *solution gas oil ratio*.

The engineer J. J. Arps had a rule of thumb; typically GOR for black oil is equal to reservoir depth divided by 10. For example, if oil was found at a reservoir depth of 7000 ft, then the amount of gas in solution would be equal to 700 cu ft per barrel (7000/10). This would not require substantial gas-handling facilities by world standards. However, some crudes can have a GOR of 3000 to 9000 cu ft per barrel or more.

Gazette – To officially announce license round offering or results, or publication of notification of acceptance of bids in official government publication (*gazette*). To gazette means to offer blocks— as in "The licenses have not been gazetted yet."

Geological horizon – A layer of rock that can be mapped. It has a distinct top and bottom. (*See* **Formation**).

Gold plating – When a company or contractor makes unreasonably large expenditures due to lack of cost cutting incentives. This kind of behavior could be encouraged where a contractor's compensation is based in part on the level of capital and operating expenditure, however, it is rare.

Government take – Government share of economic profits, typically expressed as percent. Total government share of production or gross cash flow from royalties, taxes, bonuses, profit oil.

There are a number of definitions, but the most succinct is: government take = government cash flow/gross project cash flow.

Graben – A block of rock that has dropped down (due to geologic faults) between two other blocks.

Gravity – (*see* **API gravity**)

Gravity-based structure (GBS) – Concrete production or wellhead platform fixed to the sea floor by its own weight.

Gross Benefits – In addition to the cash flow received by a government from petroleum operations, there are other benefits for a country to have foreign companies operating in-country, such as employment, added infrastructure, and technology transfer. The combination of these various economic benefits is known as the *gross benefits*.

Hard currency – Currency in which there is widespread confidence and a broad market such as that for the U.S. dollar, the British pound, Swiss franc, or Japanese yen. The opposite would be *soft currency*, where there is a thin market and the currency fluctuates erratically in value.

Heads Up – When working interest partners are paying costs and expenses in proportion to their working interest percentages, they are said to be *heads up*. When one or more partners are being *carried*, they are *not* heads up.

Hectare – Metric unit of area equal to 10,000 square meters, which also equals 2.471 acres.

High grade – A term used to describe the evaluation of acreage or a portfolio of prospects to determine which prospects or areas are best. It is used to determine which acreage to relinquish and or which prospects to drill first.

Horizon – A geological layer of rock or a formation (*see* **Formation**).

Hull formula – Compensation for expropriation in the language of many bilateral and multilateral investment treaties that states it should be "prompt, adequate and effective." This is known as the *Hull formula*. Alternate wording found in other treaties includes, "fair and equitable", "reasonable", "market value at date of expropriation," etc.

Hurdle rate – Term used in investment analysis or capital budgeting that means the required rate of return in a discounted cash flow analysis. Projects to be considered viable must at least meet the hurdle rate. Most common investment theory and practice dictates that the hurdle rate should be equal to or greater than the incremental cost of capital.

Hydrocarbon series – The various components of crude oil and natural gas composed of carbon and hydrogen atoms, i.e. the paraffin series (a subset of the hydrocarbon series):

Paraffin series (characterized by the formula C_nH_{2n+2})

C_1	-	Methane	-	CH_4
C_2	-	Ethane	-	C_2H_6
C_3	-	Propane	-	C_3H_8
C_4	-	Butanes	-	C_4H_{10}
C_5	-	Pentanes	-	C_5H_{12}
C_6	-	Hexanes	-	C_6H_{14}
C_7	-	Heptanes	-	C_7H_{16}
C_8	-	Octanes	-	C_8H_{18}
C_9	-	Nonanes	-	C_9H_{20}
C_{10}	-	Decanes	-	$C_{10}H_{22}$

et cetera

Hydrocarbon system – Proven combination of organic-rich source rocks that have been subjected to sufficient pressures and temperatures over geologic time to generate and expel hydrocarbons.

Incentives – Fiscal or contractual elements provided by host governments that make petroleum exploration or development more economically attractive. Includes such things as:

- Royalty holidays
- Tax holidays
- Tax credits
- Reduced government participation
- Lower government take
- Investment credits/uplifts
- Accelerated depreciation
- Depletion allowances
- Interest expense deductions (cost recovery)

Inconvertibility – Inability of a foreign contractor to convert payments received in soft local currency into home country or hard currency such as dollars, pounds, or yen.

Indirect tax – A tax that is levied on consumption rather than income. Examples include value-added taxes, sales taxes, or excise taxes on luxury items. (*See* **Direct Tax**)

Injection – The process of pumping gas or water in a petroleum reservoir in order to maintain pressure and enhance production.

Intangible drilling and development costs (IDCs) – Expenditures for wages, transportation, fuel, fungible supplies used in drilling and equipping wells for production.

Intangibles – All intangible assets such as goodwill, patents, trademarks, unamortized debt discounts, and deferred charges. Also, for example, for fixed assets the cost of transportation, labor, and fuel associated with construction, installation, and commissioning.

Internal rate of return (IRR) – The discount rate used in discounted cash flow analysis that yields a present value of zero for a cash flow stream is known as the internal rate of return. Rate of return (ROR) systems are based on IRR thresholds and this causes some confusion—they are called ROR systems not IRR systems.

Investment credit – A fiscal incentive where the government allows a company to recover an additional percentage of tangible capital expenditure. For example, if a contractor spent $10 MM on expenditures eligible for a 20% investment credit, then the contractor would actually be able to recover $12 MM through cost recovery (*see* **Uplift**). These incentives can be taxable. Sometimes investment credits are mistakenly referred to as *investment tax credits*.

Jack-up rig – Offshore mobile drilling vessel with a drilling rig mounted on the hull and with at least 3 tall legs through the hull. It is floated into position like a barge and hoisted above the water when the legs are mechanically lowered to the sea floor.

Joint operating agreement (JOA) – Official contract between working interest partners (members of the contractor group) in a foreign concession or PSC. The JOA outlines rights and obligations of the operator and other working interest shareholders (members of the contractor group) and means by which partners will conduct themselves. It outlines the means by which an operating committee is established, authorizations for expenditure and budgets are governed, notification deadlines, lifting rules, cash calls, and so forth.

Joint venture – The term applies to a number of partnership arrangements between individual oil companies or between a company and a host government. Typically an oil company or consortium (contractor group) carries out sole risk exploration efforts with a right to develop any discoveries made. Development and production costs then are shared pro rata between partners to the joint venture that may include the government.

Lease Option – A contractual right of an individual or a company to sign a lease, typically within a certain timeframe and upon completion of some agreed-upon work such as a feasibility study, regional study, or regional data acquisition program.

Letter of credit – An instrument or document from a bank to another party indicating that a credit has been opened in that party's favor guaranteeing payment under certain contractual conditions. The conditions are based upon a contract between the two parties. Sometimes called a *performance letter of credit*, which is issued to guarantee performance under the contract.

Letter of intent – A formal letter of agreement signed by all parties to negotiations after negotiations have been completed, outlining the basic features of the agreement, but preliminary to formal contract signing.

Levy – To impose or collect a tax or fine.

License – An arrangement between an oil company and a host government regarding a specific geographical area and petroleum operations. In more precise usage, the term applies to the development phase of a contract after a commercial discovery has been made (*see* **Permit** or **Block**).

License area – A block or concession area governed by a PSC or other type of contract between an IOC and a host government.

License splitting – A company's option to segregate a license area into segments and find partners and negotiate farm-in/farm-out arrangements for a specific segment.

Lifting – When a company takes physical and legal possession of its entitlement of crude oil, which ordinarily consists of two components under a PSC, namely: cost oil and profit oil. Lifting agreements govern the rules by which partners will lift their respective shares and how adjustments are made if a party is *over lifted* or *under lifted*.

The liftings may actually be more or less than actual entitlements, which are based on royalties, working interest percentages, and a number of other factors. If an operator or partner has taken and sold more oil than it was actually entitled to, then it is in an *over lifted* position. Conversely, if a partner has not taken as much as it was entitled to it is in an *under lifted* position. (*See* **Nomination and Entitlements**)

Lifting agreement – (*see* **Lifting**)

Limitada – Business entity that resembles a partnership with liability of all members limited to their contribution and no general partner with unlimited liability. Normally treated as a partnership by the United States for tax purposes. Similar to a Limited Liability Company in the United States, although the limitada was the forerunner.

Liquid Natural Gas (LNG) – Methane and ethane that is liquefied for shipment in specially designed refrigeration ships then re-gasified and distributed to customers through pipelines.

Liquid Petroleum Gas (LPG) – A product of distillation and contains considerably more energy than natural gas. A cubic foot of natural gas contains roughly 1000 BTUs of energy. A cubic foot of propane contains about 2500 BTUs.

London interbank offered rate (LIBOR) – The rate most creditworthy international banks that deal in Eurodollars will charge each other. Thus, LIBOR is sometimes referred to as the Eurodollar Rate. International lending is often based on LIBOR rates. For example, a country may have a loan with interest pegged at LIBOR plus 1.5%.

Loss of bargain damages – In an action for breach of contract under English common law, the plaintiff is entitled to damages so as to put him in the same position, as far as money can do it, as he would have been in if the contract has been performed.

Marginal Government take – Same as government take but with costs assumed to be zero.

Marker crude – A *marker*, or *benchmark*, *crude*, is a widely traded crude oil used as a reference for setting prices for other crudes, (e.g., Brent, West Texas Intermediate, and Dubai are benchmark crudes).

Maximum cash impairment – Maximum negative cash balance in a cash flow projection.

Nationalization – Government confiscation of the assets held by foreign companies throughout an entire industry. (*See* **Expropriation**)

Net back – Many royalty calculations are based upon gross revenues from some point of valuation, usually the wellhead, the last valve off of a production platform, or at the boundary of a field or license area. The point of sale, however, may be different than the point of valuation, and the statutory royalty calculation may allow the transportation costs from the point of valuation to the point of sale to be deducted from the actual sale price—netted back. Downstream costs between the wellhead (point of valuation) and the point of sale are sometimes referred to as net back deductions.

Nomination – Under a lifting agreement, the amount of crude oil a working interest owner is expected to lift. Each working interest partner has a specific entitlement, depending upon the level of production, royalties, their working interest, and their relative position (i.e. under lifted or over lifted), etc. Each working interest partner must notify the operator regarding the amount of oil it will lift. This is called *nominating*. Sometimes, depending upon the lifting agreement, the nomination may be more or less than the actual entitlement. (*See* **Lifting** and **Entitlements**)

Oil-in-place (OIP) – Estimates of the quantity of liquid hydrocarbons held in the pore spaces of a reservoir rock. It is understood that it is virtually impossible to recover all of the oil in a reservoir. Therefore, an estimate of the percentage of the oil-in-place that might be recovered is required to estimate recoverable reserves. (*See* **Recovery factor**)

OPEC (Organization of Petroleum Exporting Countries) – Founded in 1960 to coordinate petroleum prices of the members. Members include:

Member	Date Joined	1993 Quota (1) MBOPD	June 2000 Production MBOPD
Abu Dhabi (UAE)	1967	2,161	2,280
Algeria	1969	750	1,250
Ecuador (2)	1960		
Gabon (3)	1975	287	
Indonesia	1962	1,330	1,490
Iran	1960	3,600	3,705
Iraq	1960	400	2,565
Kuwait	1960	2,000	2,150
Libya	1962	1,390	1,420
Neutral Zone		(4)	
Nigeria	1971	1,865	2,140
Qatar	1961	378	735
Saudi Arabia	1960	8,000	8,250
Venezuela	1960	2,359	2,940
		24,520	28,925

(1) Quota did not apply for the full year

(2) Dropped out at end of 1992

(3) Dropped out at end of 1996

(4) Quotas do not apply—production shared by Saudi Arabia and Kuwait and included in their production

Operator – The company directly responsible for day-to-day operations, maintaining a lease or license, and ensuring the rights and obligations of the other members of the contractor group are met.

Operating profit (or loss) – The difference between business revenues and the associated costs and expenses exclusive of interest or other financing expenses, and extraordinary items or ancillary activities. Synonymous with net operating profit (or loss), operating income (or loss), and net operating income (or loss), economic profit (or loss) or cash flow.

OPIC (Overseas Private Investment Corporation) – A U.S. government agency founded under the Foreign Assistance Act of 1969 to administer the national investment guarantee program for investment in less developed countries (LDCs) through the issuance of insurance for risks associated with war, expropriation, and inconvertibility of payments in local currency.

Out-of-round – A term that indicates licensing of particular blocks or licenses is conducted at a time other than during an official bid round. Usually these out-of-round situations occur when companies *nominate* particular acreage that is of interest due to recent discoveries or other situations.

Over lifting – Over/under lifting is the difference between actual contractor lifting during an accounting period and the contractor entitlements based upon cost recovery and profit oil in the case of a PSC. A lifting is the actual physical volume of crude oil taken and sold.

Overspill – In international taxation, a situation where a taxpaying company has a credit for foreign taxes that is greater than its corporate tax liability in its home country so that it has an unused and/or unusable tax credit.

Pareto's law – The law of the trivial many and the critical few, is commonly known as the 80/20 rule. It has many applications and is an important analytical concept. It allows the analyst to maximize efficiency by concentrating efforts on key elements.

For example, in a portfolio of producing wells, if there is a large enough (statistically significant) sampling, 20 percent of the wells will likely produce 80% of the production. Twenty percent of the wells will represent 80% of the value. In a given basin, it is likely that 20% of the fields will hold 80% of the reserves.

Permit – In a loose sense, the term is used to describe any arrangement between a foreign contractor and a host government regarding a specific geographical area and petroleum operations. In a more precise usage, the term applies to the exploration phase of a contract before a commercial discovery has been made. (*See* **License**)

Petrophysics – The study of rock properties from either actual rock samples from the field, from coring and/or from logging methods.

Pood – Unit of measure of oil production (Azerbaijan). One pood equals 16 kilograms or roughly 62–62.5 poods per ton.

Posted price – The official government selling price of crude oil. Posted prices may or may not reflect actual market values or market prices.

Pour point – The lowest temperature at which a particular crude oil will flow. It is an indication of the wax content of the oil. Some of the famous Indonesian *waxy* crudes have pour points at nearly 100 °F.

Present value – The value now of a future payment or stream of payments based on a specified discount rate.

Price cap formulas – A fiscal mechanism where government gets all or a significant portion of revenues above a certain oil or gas price.

These formulas are typically characterized by a base price indexed to an inflation factor such as percentage change in the U.S. Producer Price Index for example. The U.S. windfall profits tax of the late 1970s and early 1980s was a variation on this theme. Malaysia and Angola have had such elements in their systems.

Prime lending rate – The interest rate on short-term loans banks charge to their most stable and creditworthy customers. The prime rate charged by major lending institutions is closely watched and is considered a benchmark by which other loans are based. For example, a less well-established company may borrow at prime plus 1%.

Produced water – Water associated with oil or gas that is produced along with the oil or gas.

Production platform – An offshore structure equipped for oil and gas production and processing. As opposed to a *wellhead platform* that is equipped for production only. Typically, production from a wellhead platform is piped to a production platform.

Production/Reserves ratio (P/R) – The percentage of total ultimate recoverable reserves produced in a peak year of production (barrels divided by barrels = %). For example, the Murcheson field in the U.K. sector of the North Sea produced at an average rate of 112,000 BOPD [40.8 MMBBLS] during 1983. With total ultimate recovery of 350 MMBBLS, this represented a P/R ratio of 11.6%.

This statistic is the inverse of the Reserves/Production ratio and/or the Reserve Life Index. These measures compare expected ultimate recovery with annual production rates (barrels divided by barrels per year = years). For example, a company has 2400 MMBBBLS of oil reserves and produces 800 MBOPD [365 billion barrels per year]. The reserve life index is 8.2 years—about average for most western oil companies. This is a slightly abstract statistic because it represents how many *years* of production the company has *if* it produces at a constant rate with no decline.

Production sharing agreement (PSA) – This is the same as a production-sharing contract (PSC). While at one time, this term was quite common, it is used less frequently now; and the term *PSC* is becoming more common—except in the former Soviet Union (FSU) where *PSA* is preferred terminology.

Production sharing contract (PSC) – A contractual agreement between a contractor and a host government whereby the contractor bears all exploration costs and risks and development and production costs in return for a stipulated share of the production resulting from this effort.

Productive horizon – A geological formation (horizon) that is known to be hydrocarbon-bearing in a given area or province.

Pro forma – Latin for *as a matter of form*. A financial projection based upon assumptions and possible events that have not occurred. For example, a financial analyst may create a consolidated balance sheet of two non-related companies to see what the combination would look like if the companies had merged. Often a cash flow projection, for discounted cash flow analysis, is referred to as a pro forma cash flow.

Profit oil – In a PSC, the share of production remaining after royalty oil and cost oil have been allocated to the appropriate parties to the contract.

Progressive taxation – Where tax rates increase as the basis to which the applied tax increases, or, where tax rates decrease as the basis decreases; the opposite of regressive taxation.

Prospect – A location where both geological and geophysical information and economic conditions indicate a feasible place to drill a well.

Prospectivity – This term deals with the exploration potential of an area and the chances for making commercial discoveries and the risks associated with exploration. An area with the potential for large discoveries and low costs and low risks would be considered highly *prospective*.

Protocol – (1) Culturally dictated forms of ceremony and etiquette that govern business relationships, meetings, and negotiations. (2) Formal document primarily used in republics of the FSU signed by parties who attend meetings or negotiations indicating various minor agreements or stages of agreement reached. These are not the same as a more formal letter of intent that usually signifies that most of the negotiations have been concluded.

Rate of return (ROR) contracts – Sometimes referred to as *Resource rent royalties (or taxes)*, *trigger taxes*, or the *World Bank Model*. The government collects a share of cash flows in excess of specified internal rate return (IRR) thresholds. The government share is calculated by accumulating negative net cash flows at the specific threshold rate of return (using a method called *compound uplifting*). Once the accumulated value becomes positive, the government takes a specified share. An example is shown as follows:

ROR	Tax Rate
0–20%	0%
20–25	30
25–30	50
> 30	70

Recoverable reserves – The hydrocarbon volumes expected to be produced economically and not left behind in the reservoir.

Recovery factor – The percentage of oil or gas in place expected to be produced. It is an estimate based upon consideration of the fluid properties such as viscosity and GOR, rock properties such as porosity and permeability, pressure gradients, well spacing, and the nature of the reservoir energy or drive mechanism.

Regressive tax – Where tax rates become lower as the basis to which the applied tax increases. Or, where tax rates increase as the basis decreases. This is the opposite of progressive taxation.

Relinquishment – This is a common contract term in exploration agreements that requires a certain percentage (often around 25%) of the original contract area be returned to the government at the end of the first phase of the exploration period. Usually additional relinquishment is required at the end of the second phase of the exploration period. Also referred to as *exclusion of areas*. Contracts typically have specific provisions for the timing and amount of relinquishment prior to entering the subsequent phases of the contract.

Reinvestment obligations – A fiscal term that requires the contractor/operator to set aside a specified percentage of profit oil or income after-tax that must be spent on domestic projects such as exploration.

Reserve life index – (*See* **Reserve/Production ratio**).

Reserve/Production ratio – This statistic compares expected ultimate recovery with annual production rate (barrels divided by barrels per year = years). For example, a company has 2400 MMBBBLS of oil reserves and produces 800 MBOPD [292 million barrels per year]. The reserves/production ratio (also called the reserve life index) is 8.2 years—about average for most western oil companies. This is a slightly abstract statistic because it represents how many *years* of production the company has *if* it produces at a constant rate with no decline.

This statistic is the inverse of the Production/Reserves ratio. (*See* **Production/Reserves ratio**).

Reserve replacement ratio – The amount of oil and gas discovered in a given period divided by the amount of production during that period.

Reservoir – A porous, permeable rock formation in which hydrocarbons have accumulated.

Reservoir pressure – The reservoir fluid pressure. (*See* **Hydrostatic Pressure**)

Resource rent tax (RRT) – Some economists refer to additional profits taxes (peculiar to the oil industry) as a resource rent tax. Australia has a specific tax based upon profits, which is referred to as resource rent tax. Normally the RRT is levied after the contractor or oil company has recouped all capital costs plus a specified return on capital that supposedly will yield a fair return on investment. (*See* **Rate of Return Contract**)

Ringfencing – A cost-center-based fiscal (or contractual) device that forces contractors or concessionaires to restrict all cost recovery and/or deductions associated with a given license (or sometimes a given field) to that particular cost center. The cost centers may be individual licenses or on a field-by-field basis.

For example, with typical ringfencing, exploration expenses in one non-producing block could not be deducted against income for tax calculation purposes in another block. Under a PSC, ringfencing acts in the same way—cost incurred in one ringfenced block cannot be *recovered* from another block outside the ringfence. Most countries use ringfencing.

Ringfencing ordinarily refers to *space* (area and/or depth), but it can also be based on *time* and *categories of costs*. It can also apply to specific reservoirs or reservoir depths and exploration vs. development expenditures.

Risk capital – Typically the drilling, seismic, signature bonuses, and costs associated with the first phase of exploration. The money placed at risk to see if hydrocarbons can be found. Often these costs have very little chance of being recovered if hydrocarbons are not found.

Royalty holiday – A form of fiscal incentive to encourage investment and particularly marginal field development. A specified period of time, in years or months, during which royalties are not payable to the government. After the holiday period the standard royalty rates are applicable. (*See* **Tax Holiday**)

Royalty leakage – In Newfoundland the *incentive* payment portion of the fees for Halliburton services which would be deductible for calculating royalty was referred to as a possible source of *leakage* i.e. it would reduce government revenue from the royalty that allowed such deductions.

Royalty oil – A percentage of the production (or revenue) paid to the mineral rights owner (government typically) free and clear of the costs of production. This represents the government oil entitlement as a result of the royalty rate in the contract between the government and the international oil company (IOC).

R factors – Some tax rates (and royalties, DMO, Gvt. Participation) are governed by pre-determined *payout* thresholds. R stands for *ratio*. Typically the contract defines R as the ratio of X divided by Y. And X is defined as *cumulative receipts* and Y is defined as *cumulative expenditures*. Cumulative expenditures include both capital as well as operating costs. When R equals 1 (one,) this is the point at which the company has achieved *payout*. Usually multiple thresholds are established. For example:

R	Tax Rate
0–1	40%
1–1.5	50
1.5–2	60
> 2	70

At the end of each accounting period, the R factor is calculated; and when a threshold is crossed, then the new tax rate would apply in the next accounting period.

Saturation – This term applies to accounting periods where there are unrecovered costs carried forward. The cost recovery mechanism is at its maximum (saturated).

Seal – An impermeable rock capable of trapping hydrocarbons in a porous reservoir rock.

Seismic – A petroleum exploration method in which acoustic (sound) energy is put into the earth with a source such as dynamite, vibrating trucks, or air guns. The sound energy reflects off subsurface rock layers and is recorded by detectors (geophones) at the earth's surface. Images of the subsurface rock layers are made with seismic surveys to locate geological structures.

Two-dimensional (2-D) seismic is where data is acquired along a single line of geophones. This has been the way data has been acquired for many years.

Three-dimensional (3-D) seismic is where data is acquired with a *grid* of multiple lines of geophones. This is a newer, more costly technology, but results have typically been quite good in terms of the quality of the data acquired.

Seismic option – A contractual arrangement between a host government and a contractor. The arrangement provides the contractor exclusive rights over a geographic area where it is obligated to shoot seismic data. After data acquisition, processing, and interpretation, the contractor has the right to enter into an additional phase of the agreement or a more formal contract with the government for the area, which usually includes a drilling commitment.

Seismic reflectors – When seismic data is acquired, there are some rocks in the subsurface that yield stronger responses *echoes* when the sound energy bounces back to the detectors (geophones) at the surface. These make it easier to *see* how the geological horizons or formations in the subsurface are folded or faulted.

Severance tax – A tax on the removal of minerals or petroleum from the ground, usually levied as a percentage of the gross value of the minerals removed. The tax can also be levied on the basis of so many cents per barrel or per million cubic feet of gas.

Shelf company – An incorporated entity, which has no assets and/or income but has gone through the process of registration and licensing. Some operations in foreign countries are started with acquisition of a shelf company because of the long delays that can be experienced setting up and incorporating a company.

Sinking fund – Money accumulated on a regular basis in a separate account for the purpose of paying off an obligation or debt.

Sliding scales – A mechanism in a fiscal system that increases effective taxes, and/or royalties based upon profitability or some proxy for profitability such as increased levels of oil or gas production (most common). Ordinarily each tranche of production is subject to a specific rate and the term *incremental sliding scale* is sometimes used to further identify this.

Example:

Typical Sliding Scale Royalty:

		Royalty
First Tranche	Up to 10,000 BOPD	5%
Second Tranche	10,001–20,000 BOPD	10%
Third Tranche	20,001–40,000 BOPD	15%
Fourth Tranche	> 40,001 BOPD	20%

Sovereign risk – Also called country risk or political risk, this term refers to the risks of doing business in a foreign country where the government may not honor its obligations or may default on commitments. Encompasses a variety of possibilities including nationalization, confiscation, expropriation, etc.

Spot market – Commodities market where oil (or other commodities) is sold for cash, and the buyer takes physical delivery immediately. Futures trades for the current month are also called spot market trades. The spot market is mostly an over-the-counter market conducted by telephone and not on the floor of an organized commodity exchange.

Spot price – Also called the *cash price*. The delivery price of a commodity traded on the spot market.

Spud – The commencement of drilling operations when a drilling rig is in-place, and a drillbit begins to penetrate the earth.

State take – The government share of profits also referred to as government take. (Although there are some consulting firms that make a distinction between government take and state take.) There are a number of definitions but the most succinct is: state take = state cash flow/gross project cash flow.

Subsidiary – A company legally separated from but controlled by a parent company who owns more than 50% of the voting shares. A subsidiary is always by definition an affiliate company. Subsidiary companies are normally taxed as profits are distributed as opposed to branch profits that are taxed as they accrue. (*See* **Affiliate**)

Sunk costs – Accumulated costs at any point in time or past costs. There are a number of categories of sunk costs:

- Tax loss carry forward (TLCF)

- Depreciation balance

- Amortization balance

- Cost recovery carry forward

These costs represent previously incurred costs that will ultimately flow through cost recovery or will be available as deductions against various taxes (if eligible).

Surrender – Surrender is often synonymous with relinquishment in the context of area reduction. However the term also is used to describe a contractor's option to withdraw from a license or contract at or after various stages in a contract. (*See* **Relinquishment**)

Swanson's Rule – A statistical method for estimating the mean of a distribution. The focus on the mean is because it is the one single value that best represents the complete distribution. And only the *means* from one distribution to another can be added. The mean is estimated by taking a *weighting* the 10th, 50th, and 90th percentile reserve estimates 30%, 40%, and 30% respectively and taking the weighted average. (*See* Fig. G–2)

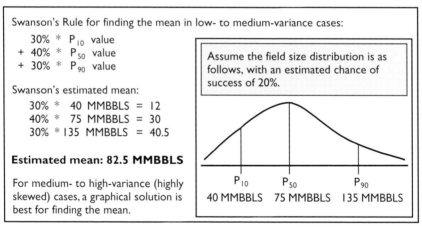

Swanson's Rule for finding the mean in low- to medium-variance cases:

30% * P₁₀ value
+ 40% * P₅₀ value
+ 30% * P₉₀ value

Assume the field size distribution is as follows, with an estimated chance of success of 20%.

Swanson's estimated mean:

30% * 40 MMBBLS = 12
40% * 75 MMBBLS = 30
30% * 135 MMBBLS = 40.5

Estimated mean: 82.5 MMBBLS

For medium- to high-variance (highly skewed) cases, a graphical solution is best for finding the mean.

P₁₀ P₅₀ P₉₀
40 MMBBLS 75 MMBBLS 135 MMBBLS

Fig. G–2 Swanson's Rule

Take-or-pay contract – A type of contract where specific quantities of gas (usually daily or annual rates) must be paid for, even if delivery is not taken. The purchaser may have the right in following years to take gas that had been paid for but not taken.

Tax – A compulsory payment pursuant to the authority of a foreign government. Fines, penalties, interest, and customs duties are not taxes.

Tax haven – A country where certain taxes are low or nonexistent, in order to increase commercial and financial activity.

Tax holiday – A form of fiscal incentive to encourage investment. A specified period of time, in years or months, during which income taxes are not payable to the government. After the holiday period, the standard tax rates apply.

Tax loss carry forward (TLCF) – In systems where expensing of pre-production costs is allowed, a negative tax base can arise, referred to as a *tax loss carry forward (TLCF)*. Also a TLCF can originate in systems where bonuses are deductible for tax calculation purposes and may be expensed.

Tax treaty – A treaty between two (bilateral) or more (multilateral) nations that lowers or abolishes withholding taxes on interest and dividends or grants creditability of income taxes to avoid double taxation.

Technical cost factor – A cost index per unit such as barrels, mcf, or BOE at some parity between oil and gas. The index is based upon the capital costs per barrel plus one-half of all operating costs per barrel. For example, if a field development is expected to cost US $300MM for 100 MMBBLS of recoverable oil, the capital costs amount to $3/BBL. If operating costs over the life of the field are expected to amount to $600 MM, then the technical cost factor would be $5/BBL. ($300 MM capital cost + $400 MM operating costs (full cycle)/2 = $500 MM *technical costs*). Technical cost factor then would be $500 MM/100 MMBBLS (or $5/BBL).

Thin capitalization rules – In countries where interest cost is recoverable or deductible, the government may introduce a backstop against the practice where overseas shareholders load the balance sheets of their in-country operations with debt, with the object of reducing host country tax exposure. Typically the government will impose an artificial (or *imputed*) capitalization structure such as 75% debt, or limit the debt/equity ratio to a certain percentage.

Tranche – Usually a quantity or percentage of oil or gas production that is subject to specific fiscal criteria. (1) The Indonesian first tranche production (FTP) of 20% means that the first 20% of production is subject to the profit oil split and taxation and this tranche of production is not available for cost recovery. (2) Sliding scale terms typically subject different levels of production (tranches) to different royalty rates, tax rates, or profit oil splits.

Example:

Typical Sliding Scale Royalty:

		Royalty
First Tranche	Up to 0,000 BOPD	5%
Second Tranche	10,001–20,000 BOPD	10%
Third Tranche	20,001–40,000 BOPD	15%
Fourth Tranche	> 40,001 BOPD	20%

Transfer pricing – Integrated oil companies must establish a price at which upstream segments of the company sell crude oil production to the downstream refining and marketing segments. This is done for the purpose of accounting and tax purposes. Where intra-firm (transfer) prices are different than established market prices, governments will force companies to use a marker price or a basket price for purposes of calculating cost oil and taxes.

Transfer pricing also refers to pricing of goods in transactions between associated companies. Often same as non-arms-length sales.

Trap – A high area on the reservoir rock where oil and/or gas can accumulate. It is overlain by a cap rock (seal).

Treaty shopping – Seeking tax benefits and treaties in various countries in order to structure an appropriately situated business entity in a given country that would take advantage of benefits that would not ordinarily be available.

Trend – The area along which a petroleum play occurs. Sometimes referred to as a fairway.

Turnover – A financial term that means gross revenues. The term is commonly used outside of the United States.

Under lifting – (*see* **Over Lifting**)

Unit-of-production depreciation – Method of depreciation for capital costs. This method attempts to match the costs with the production those costs are associated with.

Formula for unit-of-production method

$$\text{Annual depreciation} = (C - AD - S)\,\frac{P}{R}$$

Where:

C = Capital costs of equipment

AD = Accumulated depreciation

S = Salvage value

P = Barrels of oil produced during the year *

R = Recoverable reserves remaining at the beginning of the tax year

* If there is both oil and gas production associated with the capital costs being depreciated, then the gas can be converted to oil on a thermal basis.

Uplift – Common terminology for a fiscal incentive whereby the government allows the contractor to recover some additional percentage of tangible capital expenditure. For example, if a contractor spent $10 MM on eligible expenditures and the government allowed a 20% uplift, then the contractor would be able to recover $12 MM. The uplift is similar to an investment credit. However, the term often implies that all costs are eligible where the investment credit applies to certain eligible costs. The term uplift is also used at times to refer to the built-in rate of return element in a rate of return contract.

Value-added tax (VAT) – A tax that is levied at each stage of the production cycle or at the point of sale. Normally associated with consumer goods. The tax is assessed in proportion to the value added at any given stage.

Indirect taxes such as the VAT [or goods & services tax (GST)] place the company or contractor in the role of unpaid tax collector on behalf of the government. Sometimes referred to as a withholding tax.

Wildcat well – An exploratory well drilled far from any proven production. Wildcat wells involve a higher degree of risk than exploratory or development wells.

Withering Clause – (*see* **Dilution Clause**)

Withholding tax – A direct tax on a foreign corporation by a foreign government, levied on dividends or profits remitted to the parent company or to the home country, as well as interest paid on foreign loans.

Work commitment – The drilling and/or seismic data acquisition and processing obligation associated with any given phase of a PSC. This term is also used in the context of a farm-in agreement.

Working interest – The percentage interest ownership a company (or government) has in a joint venture, partnership, or consortium that bears 100% of the costs of production. The expense-bearing interests of various working interest owners during exploration, development, and production operations may change at certain stages of a contract or license. For example, a partner with a 20% working interest in a concession may be required to pay 30% of exploration costs but only a 20% share of development costs (*see* **Carried interest**). With government participation, the host government usually pays no exploration expenses but will pay its pro rata working interest share of development and operating costs and expenses.

World Bank – A bank funded by approximately 130 countries and makes loans to LDCs. The official name of the World Bank is the International Bank for Reconstruction and Development.

Abbreviations and Acronyms

$:	United States dollar
$M:	Thousands of dollars
$MM:	Millions of dollars
2-D:	Two-dimensional (as in seismic data)
3-D:	Three-dimensional (as in seismic data)
AGR:	Access to gross revenues
API:	American Petroleum Institute
APO:	After payout
B:	Billion
BBL:	Barrel (crude or condensate)—42 U.S. gallons
BCF:	Billion cubic feet of gas
BCPD:	Barrels of condensate per day
BOE:	Barrels of oil equivalent (*see* COE)

BOPD:	Barrels of oil per day
BPO:	Before payout
BTU:	British thermal unit
C:	Centigrade
Capex:	Capital expenditures
CIT:	Corporate income tax
C_1:	Methane
C_2:	Ethane
C_3:	Propane
C_4:	Butane
C_5:	Pentane
C_6+:	Hexanes plus
CNG:	Compressed natural gas
CO_2:	Carbon dioxide
COE:	Crude oil equivalent (same as BOE)
CPC:	Caspian Pipeline Consortium
C/F	Carry forward
C/R	Cost recovery
DCF:	Discounted cash flow
Dev.:	Development
DD&A:	Depreciation, depletion, and amortization
DMO:	Domestic market obligation
EBO:	Equivalent barrels of oil (same as BOE and COE)
EV:	Expected value (same as EMV)
EMV	Expected monetary value (same as EV)

ERR: Effective royalty rate

F: Fahrenheit

ft: Feet

FVF: Formation volume factor

GOR: Gas oil ratio (typically cubic feet/BBL)

Gvt. Government

G&A: General and Administrative (as in costs)

H_2S: Hydrogen sulfide

$I-C_4$: Iso-butane

$I-C_5$: Iso-pentane

IRR: Internal rate of return

JOA: Joint operating agreement (same as JVOA)

JV: Joint venture

JVOA: Joint venture operating agreement (same as JOA)

KCS: KazakhstanCaspieShelf

KE-1: Kashagan East-1 well

KE-2: Kashagan East-2 well

KW-1: Kashagan West-1 well

LGR: Liquid gas ratio (typically in BBLS/MMCF)

LNG: Liquefied natural gas

LPG: Liquid petroleum gas

m: Meters

M: Thousand

MBBLS: Thousand barrels

MCF: Thousands of cubic feet (gas)

MCFD: Thousand cubic feet of gas per day

MM: Million

MMBBLS: Million barrels

MMBOE: Million barrels of (crude) oil equivalent

MMCF: Million cubic feet of gas

MMCFD: Million cubic feet of gas per day

MMT/yr: Millions of (metric) tons per tear

mol: Molecular

mol vol: Molecular volume

% mol vol: Percent molecular volume

mw: Megawatt

$N-C_4$: Normal butane

$N-C_5$: Normal pentane

N_2: Nitrogen (molecular nitrogen)

NGL: Natural gas liquids

N/A: Not applicable or not available

No.: Number

NPV: Net present value (same as PV)

OKIOC: Offshore Kazakhstan International Operating Company

Opex: Operating expenses

PSA: Production sharing agreement (same as PSC)

PSC: Production sharing contract (same as PSA)

psi: Pounds per square inch

P&A: Plugged and abandoned

PV: Present value (same as NPV)

P/O: Profit oil

P/R: Production to reserves ratio

P_{50}:   Probabilistic reserve figure—50th percentile

R factor: *R* here stands for "ratio" (*see* Glossary)

RLI: Reserve life index

ROR   Rate of return (sometimes refers to IRR)

SLD: Straight-line decline

STOIIP: Stock tank oil initially in place

STOIP: Stock tank oil in place

TLCF: Tax loss carry forward

t/d: Tons/day

TCF: Trillion cubic feet of gas

TFE: TotalFinaElf

UAE: United Arab Emirates

VAT: Value added tax

Vs.:   versus (Latin)

Wt%:   Weight percent

Z: Gas compressibility factor

°:   degree

Index

A

B

C

D

E

Energy Policy and Conservation Act of 1975, 260
Entitlement barrels, 251, 266-268
Entitlement index, 74, 102-103, 173-174: calculation, 102-103
Equatorial Guinea, 50, 294
Equity split, 59
Estimating problems, 86
Ethane, 126
Expectations, 220, 222
Expected monetary value (EMV), 7, 109-110: model, 7
Expected ultimate recovery, 272-274
Expected value theory, 2, 7
Expected value formula, 159, 163-164
Expected value (EV), 2, 7, 109-110, 159, 163-164, 225-227, 229-233: theory, 2, 7; model, 7; formula, 159, 163-164
Exploration agreement, 41
Exploration and development production-sharing agreements (EDPSA), 1
Exploration costs, 94, 179-180, 262: well costs, 179-180
Exploration production-sharing agreements (EPSA), 1
Exploration rights, 3, 190-191: and development, 3
Exploration well costs, 179-180
Export tariffs, 89

F

Factor R-based systems, 42-44, 154-155, 209-210: R thresholds, 42-43; payout/yield, 43; royalty rate, 43-44
Farm-in strategy, 84-85
FAS 19, 259-261
FAS 25, 261
FAS 69, 260-263
Fast-track approach, 85
Federal Securities Laws, 260

G

H

I

M

N

O

Q

R

Rate of return (ROR) systems, 17-18, 45-51, 154-155, 210: example, 45-51

Rate sensitivity (production), 144-146: water-drive reservoirs, 145-146; solution gas-drive reservoirs, 146; gravity-drainage reservoirs, 146

Ratio factor. SEE R factor-based systems.

Reasonable certainty concept, 265

Recoverable gas vs. oil, 118

Recovery of cost (R/C) contract, 311

Recovery of costs, 4, 32, 59, 67, 72, 101, 141, 154, 174, 212, 311: cost recovery limit, 32, 67, 72, 101, 154, 174; contract, 311

Regressive fiscal system, 15-18

Relative economics, 61

Relevance (government take), 244-246

Relinquishment, 153, 192-193

Rent theory, 2

Rental payments (acreage), 58, 72

Reserve recognition accounting (RRA), 101-102, 259-264: definitions, 263-264

Reserve replacement, 110-113

Reserve value estimates, 268-269

Reserve values, 27, 247-255, 268-269: value in the ground, 247-255; estimates, 268-269

Reserves disclosure criteria, 263

Reserves estimates, 247-269: value in the ground, 247-255; rules of thumb, 255-269

Reserves, 101-102, 110-113, 173-174, 183-184, 194-195, 214-215, 223-224, 247-269: values, 27, 247-255, 268-269; reserve recognition accounting, 101-102, 259-264; replacement, 110-113; estimates, 247-269; disclosure criteria, 263; value estimates, 268-269

Reservoir characteristics, 205-206, 272-273: depth/pressure, 205-206; damage, 272

Reservoir damage, 272

S

T

U

V

W

X-Z